A RESEARCH ANNUAL

RESEARCH IN THE HISTORY OF ECONOMIC THOUGHT AND METHODOLOGY

Series Editor: Warren J. Samuels

RESEARCH IN THE HISTORY OF ECONOMIC THOUGHT AND
METHODOLOGY   VOLUME 19-A

A RESEARCH ANNUAL

EDITED BY

WARREN J. SAMUELS and JEFF E. BIDDLE

Department of Economics, Michigan State University,
East Lansing, MI 48824, USA

2001

JAI
An Imprint of Elsevier Science
Amsterdam – London – New York – Oxford – Paris – Shannon – Tokyo

ELSEVIER SCIENCE B.V.
Sara Burgerhartstraat 25
P.O. Box 211, 1000 AE Amsterdam, The Netherlands

First edition 2001

Library of Congress Cataloging in Publication Data
A catalog record from the Library of Congress has been applied for.

ISBN: 0-7623-0703-X
ISSN: 0743-4154 (Series)

♾ The paper used in this publication meets the requirements of ANSI/NISO Z39.48-1992 (Permanence of Paper).
Printed in The Netherlands.

CONTENTS

vi

LIST OF CONTRIBUTORS

Mark Blaug	Faculty of Economics, University of Amsterdam
Y. S. Brenner	Faculty of Social Science, University of Utrecht
Anthony Brewer	Department of Economics, University of Bristol
Tyler Cowen	Department of Economics, George Mason University
Paul Diesing	Department of Political Science, SUNY, Buffalo
Soltan Dzarasov	Department of Economic Theory and Entrepreneurship, Russian Academy of Sciences
Ross B. Emmett	Department of Economics, Augustana University College
Luca Fiorito	Department of Economics, New School for Social Research
Samuel Hollander	Department of Economics, Ben Gurion University of the Negev
Glenn Hueckel	Department of Economics, Purdue University
S. Todd Lowry	Department of Economics, Washington and Lee University
Kirsten K. Madden	Department of Economics, Millersville University
Thomas Mayer	Department of Economics, University of California-Davis
Perry Merhling	Department of Economics, Barnard College

Kenneth Mischel	Department of Economics and Finance, Baruch College, CUNY
Spencer J. Pack	Department of Economics, Connecticut College
Melville W. Reder	Graduate School of Business, University of Chicago
Murray Wolfson	Department of Economics, California State University, Fullerton
Jeffrey T. Young	Department of Economics, St. Lawrence University
Stephen Thomas Ziliak	Department of Economics, Bowling Green University, Emory University

EDITORIAL BOARD

ACKNOWLEDGMENTS

The editors wish to express their gratitude for assistance in the review process and other consultation to the members of the editorial board and to the following persons:

Philippe Broda	William Kern
Jose Luis Cardoso	Philippe Le Gall
Richard Dawson	Steven Medema
Ken Dennis	Sandra Peart
Paul Diesing	Esther-Mirjam Sent
Kevin Hoover	E. Roy Weintraub
Niels Kaegard	Milan Zafirowski

METHOD CHOICE IN THE HISTORY OF ECONOMETRICS: A DECISION CRITERIA TAXONOMY AND ANALYSIS OF THE REPORT OF ECONOMETRICIANS' DECISION MAKING ACROSS SOURCE MATERIALS

Kirsten K. Madden

ABSTRACT

This research considers scientific decision making, providing the first systematic study and explicit description of decision criteria employed by econometricians in method choice decisions. The documentation of econometric history is used as the source of information regarding method choice. Information from the histories is used to create a method choice taxonomy and a method choice database. Contingency tables are formed from the database and chi-square tests are run on hypotheses. The taxonomy reveals that econometricians base their method choice decisions on scientific, pragmatic, theoretical, metaphysical, and sociological factors. Statistical results suggest that econometricians' reports of the

Research in the History of Economic Thought and Methodology, Volume 19A, pages 1–31.

ISBN: 0-7623-0703-X

reasons for method choice vary depending upon the type of source in which the information is presented.

I. INTRODUCTION

The analysis of choice is the basis for much research in the field of economics. Economists, for instance, may study the choice between consumption goods, production inputs or the alternative uses of time of an economic agent, whereas historians of economic thought may concentrate on theory choice. A related and less studied concern is method choice, i.e. the selection of empirical tools and analytical techniques by economists. Unlike work in economic theory, research on issues of choice in economic methodology has not provided a cohesive, systematic framework with which to analyze the selection process as it relates to methodological concerns.

This chapter addresses two major methodological issues arising in the history of econometrics. First, what factors are involved in the econometrician's selection of empirical tools and analytical techniques? This question may be interpreted as discovering and defining the decision criteria x, y and z used by econometricians in:

$$\text{method choice} = f(x, y, z, \ldots).$$

The objectives are to determine the potential decision criteria which may influence an econometrician's decision to choose a particular method in econometric research and to systematically, coherently organize them into a comprehensive method choice taxonomy.

The literature on the history of econometric thought has been expanding throughout the past decade. The problems addressed in the current body of literature about the history of econometric thought suggest an implicit concern with the issue of method choice. Consider some of the titles of works in the history of econometrics: "Advances in Confluence Analysis and the Decline of Bunch Maps" (Hendry & Morgan, 1989), "The Fall of OLS in Structural Estimation" (Epstein, 1989), "Rejection of Maximum Likelihood Estimation" (Epstein, 1987), and "Haavelmo's Probability Model: The New Consensus" (Morgan, 1990). The commonality in these works is their concern with method selection and in the diffusion process peculiar to particular methods. Unfortunately, the specific cases given in the histories do not provide a means for understanding the general nature of the process. The lack of a framework for analyzing method choice lends an arbitrary air to the existing historical discussions of individual method choice decisions because there is no underlying support structure to the analysis. This study supplies this support structure, providing the basis for consistency and depth in discussions regarding method choice in the history of econometrics. A primary result of this study is that historically, econometricians have been influenced by scientific, pragmatic, theoretical, metaphysical and sociological factors in their scientific decision making.

Given the variety and complexity of reasons underlying method choice in the history of econometrics, the second methodological question addressed in this work concerns econometricians' reports of their decision making in journal articles, books and informal accounts of econometric research. Do these different sources of information typically yield different criteria concerning method choice? In the second part of this study, the method choice framework is applied to analyze the reporting of econometric decision making. A primary result is that there is inconsistency in the reporting of the method choice criteria in econometricians' documentation of their scientific research.

II. METHOD

Two terms vital to this work require explicit definition due to the ambiguity of their meanings in various contexts. First, "method" refers to "a systematic procedure or mode of inquiry;" the terms "tool," "technique," and "method" are used interchangeably.[1] Second, "choice" refers to "the act of choosing" or "selection."

This study is an inductive inquiry to develop a framework for the analysis of method choice in econometrics. An exhaustive analysis of the literature in the history of econometrics was undertaken to identify the potential factors which influence the acceptance or dismissal of various econometric methods. Six primary sources of information constitute the foundation for this study. Two of these sources are books about the history of econometrics: Epstein (1987) covers the history of structural estimation, and Morgan (1990) covers the early history of econometrics from the late nineteenth century to the 1940s. Christ (1952) constitutes the third source of information for this study; it is a detailed report of the first twenty years of the Cowles Commission. The remaining three sources of information are articles and include a piece by Hendry and Morgan (1989) about the history of confluence analysis, a history of ordinary least squares presented in Epstein (1989), and an article on the evolution of the econometric approach at the London School of Economics, written by Christopher Gilbert (1989).

Each of these histories was analyzed for criteria which were used to choose methods by researchers in the history of econometrics. These decision criteria were extracted to create a "data set" of citations from the historians of econometrics in which each quote suggests a reason for choosing or rejecting a particular method by some decision maker.[2] Approximately 53% of the observations are derived from Epstein (1987), 36% from Morgan, 1% from Christ, 2% from Hendry and Morgan, 3% from Epstein (1989) and 4% from Gilbert.

These citations were organized into a coherent framework of method choice categories. The categorization process was iterative in nature. Citations were initially grouped together; categorical definitions were determined by extracting similar characteristics from the grouped citations. The citations were then reviewed for consistency with the category definitions and vice versa. Revisions included adding and deleting citations, changing classification definitions, and reviewing the categories with respect to one another to minimize definitional overlap. In a taxonomy of this kind, there is inevitably some inter-category overlap. As the categories for method choice were placed in the taxonomy, the categorical definitions were reviewed in relation to one another. The goals of these reviews were to make sure that no category was a subset of another and to minimize the amount of overlap across categories.[3] The results of this procedure are detailed in section III.

As described above, this study generally relies on the writings of the historians, rather than on the original research material of the econometricians. There are compelling reasons for this reliance on historical accounts, the most important being that the decision criteria employed for the selection of particular techniques are frequently implicit (or left out) in the original sources. The works of the historians are particularly valuable due to their ability to make explicit the factors which econometricians, who were actively involved in the decision-making process, may not have clearly perceived or described. Using information from econometric history, this study analyzes the method choices of *econometricians in econometric history*. Information obtained from the histories of econometrics provide an approximation to this decision making process. The results of this study provide a framework for further methodological analysis. Using the method choice taxonomy in this work, the next step in this research agenda is to analyze method choice through new sources of information obtained through surveys of and interviews with econometricians as well as review of their primary materials documenting econometric activity.

Concerns of bias in this study due to the use of secondary sources remain. Madden (1995) provides the results of a detailed descriptive and statistical analysis of historian bias in the reporting of method choice criteria. In general, each historian is found to have their own predilections in terms of their emphases regarding method choice. This result is important because it reinforces the need to undertake this type of methodological study *across* historians, rather than to focus on a single historian's work. This study minimizes the impact of this bias by combining results from six histories in econometrics. Given the high frequency of observations accruing from Epstein, the primary problem in this study resulting from historian bias is that the numbers are likely to be too high in certain method choice categories, particularly the technical reasons, defined below. In general, limited historical review of the sociological aspects of econometric history may be partially responsible for the low numbers accruing to the sociological category.

The procedures employed in the fourth section of this chapter rely upon the development of a method choice database and the application of statistical methods to the database.[4] The database is based upon the extracted citations from the historical literature. It incorporates the method choice taxonomy as well as such information as the name of the decision maker, the method considered, the source of information in which the historian found the facts for their histories, and the date of the decision. For each citation, an attempt was made to extract information for each of the variables defined in the database. The result is that each record (row) in the database consists of one decision criterion used by a predefined decision maker to make a decision regarding one particular method. The database contains 32 variables and 1,181 observations, of which 829 are by "individual" decision makers.[5]

Ninety-five individual decision makers are responsible for the 829 method choice reasons in the database. The five decision makers most frequently cited and the approximate percentage of the total observations accounted for by these five are: Jan Tinbergen and Ragnar Frisch (8% each), Tjalling Koopmans (5.5%), and Henry B. Mann and Abraham Wald (4% each). The next four most frequently mentioned economists, responsible for a total of 15% of the database observations, are Trygve Haavelmo, Henry Schultz, Lawrence Klein and Henry L. Moore.[6] The observations in the database fall primarily in the forty year time period 1920–1959; the percentage of observations over decade from 1860 to 1989 are provided in Appendix B.

In undertaking the methodological analysis, the question of interest is stated as an explicit hypothesis directly relevant to the information in the database. The information in the database is then organized into contingency tables, with the method choice taxonomy considered in relationship to some other pre-defined database variable. The frequency of citation for each decision criteria is then tabulated, and these frequencies provided the numerical foundation for the ensuing analysis. The information in the database was thus utilized to test the hypothesis, and these results were fundamental in shaping the final conclusions.

III. ANALYTICAL FRAMEWORK FOR THE STUDY OF METHOD CHOICE

The following outline provides a summary of the categories which historically have affected the choice of method in econometrics. A brief definition is provided for each of the method choice categories in the remainder of this section.

A. Science and Method Choice

A1. Positivism/Scientific Paradigm
A2. Technical Reasons
A3. Empirical Results

B. Pragmatism and Method Choice

B1. Data-Related Issues
B2. Computation Costs
B3. Difficulty in Understanding Method
B4. Subjective Pragmatism

C. Suitability and Method Choice

C1. Method/Theory Correspondence
C2. Range of Economic Problems
C3. Underlying Assumptions of Method

D. Metaphysics and Method Choice

D1. Underlying Philosophy/Conceptualization
D2. Methodology
D3. Aesthetics

E. Sociology and Method Choice

E1. Background/Training
E2. Specific Individuals Involved/Acquaintances
E3. Methods and Marketing
E4. Interdisciplinary Communication

CATEGORY DEFINITIONS

A. Science and Method Choice

Econometricians use differing combinations of economic theories, mathematics and statistics, and empirical evidence to substantiate their economic ideas. In undertaking econometric research, certain predefined rules and procedures must be followed. The positivist or scientific paradigm provides general principles to guide econometric activity, including the issue of method choice. Within this paradigm, econometricians have explicit technical requirements with which to appraise the methods under consideration. Finally, when they follow the rules, econometricians obtain empirical results which can be used to evaluate their method choice decisions.

A1. Positivism/Scientific Paradigm

The ideal practice in science is to develop a hypothesis and objectively test the hypothesis in a setting in which all variables are controlled for except those of specific interest. Unfortunately, this ideal cannot be strictly adhered to in much scientific research, including econometrics. Since by nature econometric practice falls short of the scientific paradigm, issues have been raised regarding the ability of particular methods to handle the anomalies of undertaking statistical research in the social science setting. With respect to this category, the two primary discussions regarding method choice include the inability to perform controlled experiments and the issue of objectivity. Techniques which are applicable beyond the realm of the scientific laboratory are obviously preferable in econometrics. In contrast, the subjectivity involved in the application of particular methods has provided grounds for dismissal of these methods.

A2. Technical Reasons

Statistical criteria are frequently cited as factors affecting method choice in econometrics. Methods may be accepted or rejected based on their statistical properties: bias (in finite samples or asymptotically), consistency, and efficiency. Mathematical and algebraic concerns have also been voiced with regard to method choice. Issues include the possibility of convergence to maxima or minima that are not global and the inability of a method to provide unique solutions. The issue of stability, or the tendency to return to equilibrium, arises in discussions about the selection of econometric models. The identification issue (or whether or not information on the structural parameters of a system can be recovered) in simultaneous equations systems is also raised repeatedly in the history of econometrics. Finally, econometricians are also interested in the ability of methods to capture the effects of omitted independent variables. If a method is not reliable in such a scenario, it may be rejected.

A3. Empirical Results

Methods are often evaluated based upon a consideration of their empirical results. This category for method choice focuses on the forecasting capabilities of methods, on the empirical results obtained from methods, and in general on the empirical insights methods provide into economic considerations.

B. Pragmatism and Method Choice

This class of method choice categories focuses on the practical affairs in scientific activity. Econometricians have problems to solve and goals to obtain, material to work with and resources to use. At some point, the mundane issues of whether a method will work and actually making it work must be addressed. A variety of questions may be raised in the course of making an econometric project operational. Is the method practical from a resource standpoint? That is, can it work on the materials (e.g. data) provided, and will the method provide concrete results given the inevitable constraint of resources (time, machine and mental power) econometricians must face? From the perspective of the researcher, is the method itself accessible and feasible to use? Does the researcher believe that the method is an effective means of obtaining his goals? These sorts of issues form the foundation for this section.

B1. Data-Related Issues

Features of economic data have provided the basis for decisions to accept or reject a variety of methods. Issues regarding the quantity of economic data and its quality come up frequently, as do concerns regarding the applicability of a method to data characteristics. Related to the issue of the quantity of data is the problem of structural change or the lack of long runs of homogeneous data in economics. The requirement for long runs of stationary data may be problematic for some methods. Other problems include the applicability of a method to data characteristics such as time-related data or autocorrelation, sensitivity to seasonality and trend components in the data and inefficient utilization of data information. Finally, some method choice decisions are based upon the usefulness of a particular tool to investigate economic data; a method may be chosen simply if it allows the researcher to "gain more familiarity with the data at hand" (Epstein, 1987, p. 101).

B2. Computation Costs

The focus of this category centers around the costs incurred in econometric work due to the application of a particular method. Computation costs concern the "computational job" (Epstein, 1989, p. 102), the "computational labor" (p. 103) the amount of "'labor involved'" (Morgan, 1990, p. 147) or the "time-consuming nature of doing" a method (p. 139). The costs of using a method on economic data provided many constraints to econometric work prior to the computer revolution. A method might be rejected if its practical application was extraordinarily costly, "tedious" (p. 106) "cumbersome" (p. 59) or "entailed a crushing burden of arithmetic" (Epstein, 1987, p. 39); a method would often be accepted if the estimates resulting from use of the method were "easily computable" (p. 151), "quicker" (Morgan, 1990, p. 88), or "more economical" (Christ, 1952, p. 39) compared to alternative tools. In some respects, the computer revolution has made this category less of an issue for some tools than in the past, although clearly computation costs continue to play a role in tool choice in econometrics. Particularly increases in computability may be related to increased innovation and use of more calculation-intensive techniques.[7]

B3. Difficulty in Understanding Method

As with the category which follows, the pragmatic concern "difficulty in understanding a method" is subjective and relative to the econometrician; what is difficult to the technically inept may be elementary to those more technically inclined. Also of relevance for method choice in this category is the interpretability of a tool's results. For example, with regard to the periodogram, Morgan states that there were "difficulties of interpreting the periodogram results", or in other words, the "method seemed to produce too many cycles to allow sensible economic interpretation" (p. 36). Difficulties may be faced in understanding the technique itself or in evaluating the results obtained from the method, and these difficulties, if insurmountable, may lead a researcher to abandon the use of the method under consideration.

B4. Subjective Pragmatism

In some cases, the choice of a tool by an individual researcher is based primarily on that practitioner's subjective valuation of the tool's usefulness for prescriptive purposes. Certain econometric tools have been accepted based on beliefs in their forecasting potential, or on their ability to aid in projects related to "social engineering", or generally based on a researcher's conviction that a method may further a particular political agenda. The important concept to note in this category is that tools are used or rejected based on the *belief* by the researcher that their use is pragmatic for the researcher; this is a far different kind of pragmatism than, for instance, rejecting a method due to data constraints.

C. Suitability and Method Choice

The issue in this section is whether or not a method is suitable to task. There is an economic issue to address and methods to use. Researchers are concerned with the applicability of a method to the problem of interest. Do the methods correspond to the economic theory underlying the task? Do the assumptions underlying the method hold for the problem of interest? Does the method apply to the problem at all or is it fundamentally unsuitable for the issue?

C1. Method/Theory Correspondence

The concern of this decision criterion for method choice regards the relationship between a method and economic theory. A broad variety of views hold regarding the ideal method/theory relationship, and these views may be placed along a method/theory correspondence continuum. At one extreme, a method strictly adheres to the theory of interest. At the opposite pole, a method is void of economic content. Between these extremes lay concerns that methods provide some representation of theory or that methods are in some way interpretable in terms of economics.

C2. Range of Economic Problems

If a method has a limited range of questions to which it can be applied, then, whatever the reason, if the questions become uninteresting the method ceases to be used. In contrast, if the range of relevant questions increases for a particular method, then one would tend to see increased usage of the method. Thus, the expansion of simultaneous equations models to microeconomic applications led to continued usage of this tool by applying it to a broadened range of problems

C3. Underlying Assumptions of Method

The assumptions underlying a method may be reviewed for applicability to the economic problem being addressed. If the assumptions are implausible, this may provide material with which to reject a method. Examples of such problematic assumptions are wide and varied in econometric work. For example, in early demand studies assumptions were made on the shifts in the curves including the independence of shifts, and the equality of shifts, elasticity assumptions were made and linearity was frequently assumed. In macroeconometric models of the late 1930s and 1940s assumptions were made concerning whether the models were correctly specified, model stationarity may have been assumed, and identifying assumptions were made. As econometrics has evolved, more abstract assumptions have become the norm. For example, assumptions are now couched in terms of specific distributional forms of the disturbance terms, error variance matrix restrictions, and exogeneity assumptions. Although these assumptions and the tests for them have acquired a more technical flavor, their relationship to economic theory is still a fundamental issue in method choice.

D. Metaphysics and Method Choice
In choosing the term "metaphysics" to describe one aspect of factors affecting method choice decisions, I am not alluding to any sort of mystical or spiritual consideration; the spirit of econometrics past is not frequenting the subconscious of Dr. Scrooge suggesting that she "do it this way". For the purposes of this work, metaphysics is the "study of what is outside objective experience" and also as "the system of principles underlying a particular study or subject" (*Webster*, 1983). Although these two definitions illuminate different aspects, they both have the effect of shifting the focus of this study away from concrete reality to the abstract and more subjective nature underlying econometric activity. Jointly the two definitions encompass the three sub-categories of metaphysics: underlying philosophy/conceptualization, methodology, and aesthetics.

D1. Underlying Philosophy/Conceptualization
The common thread within this category is that the acceptability of any method is based on the *Weltanschauung* of the researcher, in both a broad and a narrow sense. The way in which economic life and man's behavior is conceptualized has repercussions for method choice. Different methods will be used based on whether the economic process is viewed "as a largely deterministic exercise" or whether it "involve[s] elements of chance" (Morgan, p. 241). Choices or beliefs regarding economic theory also have ramifications for method choice. The emphasis of this category is that when theory is questioned or rejected, this decision will have ramifications for method choice. For example, the changing conceptualization of the business cycle (from crises to periodic and constant events to the idea that cycles are only "roughly regular" occurrences) had major ramifications for method choice in this field. The periodogram method is applicable to business cycle research if the cycles are periodic and have constant amplitude, but not otherwise. As another example, a researcher rejecting theories based on "equilibrium notions" in economics believed this to be grounds for rejecting simultaneous equations models in econometrics.

D2. Methodology

The methodology of science pertains to the criteria scientists rely upon when undertaking an investigation. *Webster's Ninth New Collegiate Dictionary* defines this set of criteria as any "body of methods, rules and postulates" employed by a discipline. In some cases, the method choice hinges upon the methodological rules underlying the study. The specific rules which have been discussed in the historical literature and which have affected method choice include the methodological scope of research, inductive versus deductive inquiry, and, in general, the specific rules to be followed in a particular study. With regard to the issue of scope, the role for which a technique is found acceptable limits the use of a method. For example, in the early period of econometric history, the role of econometrics was limited to measurement; tools which are primarily applicable for theory testing or theory development were frequently rejected (Morgan, 1990).

D3. Aesthetics

Aesthetic considerations concern the qualitative, immeasurable characteristics of an object being scrutinized. The object under investigation may be a piece of art, a film, a mathematical proof, a theory, or even an econometric technique. Consider the stereotypical example of a mathematician raving over the elegance of a particular proof; it is not that the statement has been proven, but rather the style in which the solution has been obtained which so intrigues the mathematician. Qualitative characteristics have been noted in a paper on theory choice as well. In "Theory Choice in Economics", the authors ask "whether other, non-empirical criteria for judging the acceptability of theories may be invoked," and they proceed to develop a set of criteria (Tarascio and Caldwell, 1979, p. 991). Their set includes logical consistency, elegance, extensibility, generality, theoretical support or multiple connectedness, fertility fruitfulness and heuristic value, and simplicity. I would like to expropriate this list as it relates to method choice and add the terms clarity, concision, coherence, cohesiveness, and consistency. In short, by "aesthetics" I refer to all those qualitative tendencies that scientists have historically consulted when deciding upon the preferability of one proof, theory, or method over another proof, theory, or method.

E. Sociology and Method Choice

This section incorporates the social determinants of method choice. The relevant issues include factors influencing what the researcher has learned, who the researcher knows, the functions involved in marketing a method, and the transmission of techniques across disciplinary boundaries. Each of these issues are amplified below in the categories background/training, individuals involved/acquaintances, methods and marketing, and interdisciplinary communication.

E1. Background/Training

The background and the type of training a researcher receives influences the methods chosen by that person. Course availability, textbook and journal coverage, the amount and type of technical material one is exposed to, and even previous training in other fields are some of the factors which influence the education of a researcher and may impact method choice. A primary form of knowledge transmission in the economic discipline occurs through the provision of classes in universities and through the written resources relied upon in the courses. Both of these media provide outlets for the diffusion of methods to future researchers. With regard to the exposure to technical material, historically those exposed to mathematics and statistics have been more open to the development and application of mathematical and statistical tools. The influence of a mentor may also have an influence on the choice of methods by the "disciples."

E2. Specific Individuals Involved/Acquaintances

In general, the influence, personalities and interests of people involved shape the type of research approaches undertaken and ultimately the method choice decisions. With regard to acquaintances as a factor influencing method choice, a decision may be based upon who you know and allow to influence your decisions. The idea behind the "who you know" aspect of this category is that perhaps the method choice decision would have been different had a specific person not been the acquaintance of the researcher making the decision. The focus here is on interpersonal influence. This influence may be given through purely informal media as, for example, in conversation, memoranda and correspondence, or it may be provided in more formal media such as in the provision of a journal article to a researcher.

E3. Methods and Marketing

Discussions about method choice suggest that certain conditions are needed for individual acceptance and, in general, for the diffusion of methods across the economics profession. *Webster* defines the term marketing as "an aggregate of functions involved in moving goods from producer to consumer". If econometric techniques are equated with the word "goods", the producers being the researchers involved in the development of the technique for econometric use and the consumers being the economists who may potentially use the method, then an illuminating analogy can be developed to understand one form in which social aspects influence method choice. Some of the functions involved in moving these techniques from developer to user which have ramifications for the likelihood of method acceptance and eventual diffusion include: (1) resource commitment; (2) the existence of sellers; (3) an appropriate marketing strategy; (4) the existence of a potential market; and (5) the extent of the market. With regard to marketing strategy, four notable factors involved in marketing a technique are: (a) the style in which research about methods is communicated; (b) method endorsement; (c) where the work is presented; and (d) when the research is presented. Finally, journals have the potential to regulate the market and thus impact the diffusion of a method. For example, (Epstein, 1987) notes that "econometricians in that era were already under pressure from the journals to avoid complicated mathematics" in the first few decades of this century (p. 24).

E4. Interdisciplinary Communication
Communication and cooperation across disciplinary boundaries contribute to method choice. The historical importance of this type of communication for econometrics lies in the fact that many of the econometric tools have their origins in other disciplines such as, for example, biometrics, statistics and mathematics.

This closes the detailed examination of the decision criteria involved in method choice in the history of econometrics. These decision criteria have been catalogued into five different general groups which range from scientific to sociological factors. As is apparent from such a broad range of criteria, the evidence suggests that method choice has followed no simple formula throughout the history of econometrics.

IV. ANALYSIS OF THE METHOD CHOICE TAXONOMY WITH THE SOURCE OF INFORMATION

The historians of econometric thought used a variety of sources to piece together their stories. These sources include journal articles, books, correspondence and interviews to name only a few. In this section, analysis turns from consideration of the decision making process itself to questions concerning the documentation and report of scientific decision making. The purpose of this section is twofold. First and more generally, do different sources of information yield different types of information? More specifically, are the frequencies of observations attained in the method choice decision criteria dependent on the source of information from which the observation was derived? If the answer to this question is yes, the second function of this section is to determine which sources of information are dominant for particular decision criteria categories.

The impetus for this particular study is the work undertaken by Knorr-Cetina (1981) concerning the manufacture of scientific knowledge. Knorr-Cetina states that as a report of scientific activity, the journal article "hides more than it tells on its tame and civilised surface". She shows that the scientific paper "deliberately forgets much of what happened in the laboratory, although it purports to present a 'report' of that research" (p. 94). In particular, Knorr-Cetina argues that scientific procedures are highlighted in the official reports of research results while sociological aspects of research activity are systematically polished out of scientific papers. As mentioned above, the historians used a number of sources of information in their accounts on the history of econometrics. Although distributed across the various types of sources, the observations in the database are not equally distributed over these sources. In fact, journals are the most heavily relied upon source, providing almost one half of the information for this study. With just over one quarter of the 829 observations in the individuals database derived from books, they are the second most frequently cited source of information on method choice. Informal sources such as memoranda, correspondence and minutes from meetings, are the foundation for 15% of the information on method choice; 9% of the information for this study was not explicitly tied to a particular source. Therefore, the remainder of this section is devoted to studying the implications of the source of information on the frequency of observations for the method choice categories.

Statistical testing (to be described later) of the information divided by source type suggests that differences between the source types are apparent, but not the ones I envisioned. Conceptually it seemed that books would have much in common with journals because both sources are polished accounts which require approval by external parties in order to be published. Thus, I viewed a dichotomy based on formal versus informal sources of information, with books and journals in the former and unpublished materials in the latter. It turns out that the information derived from books does not share the same characteristics as that obtained from journal sources. The book information actually has more in common with the information derived from informal sources. This finding will be amplified on below. The implication is that no relevant insights come from the analysis of a formal/informal breakdown. The division of interest is between the three groupings of journals, books and informal sources as well as between journals and the combination of books and informal sources.

Table 1 presents a list of all the sources of information from which are derived the quotes on method choice. These sources are broken down into three

Table 1.   Sources of Information on Method Choice.

Group A

books
dissertations
monographs

Group B

essays in books
research reports
journal articles
book reviews
journal article replies to criticisms
journal article replies to replies
journal article debates on methods

Group C

archival information
answers to federal questionnaires
lecture notes
research plan documents
unpublished drafts
unpublished book reviews
memoranda
minutes from meetings
correspondence
conference papers
discussion papers
privately circulated unpublished mimeographs
interviews
historical sources

groups. Group A includes lengthy published accounts of economic research. Group B consists of more concisely stated publications in economics which generally follow journal-like norms in publication. Group C combines primarily unpublished, and thus less polished, accounts of research activity. These three groups will be utilized in different combinations to test hypotheses on the differences among decision criteria with respect to source type.

The chi-square is the statistical test used to determine whether the same kind of information about method choice is present in these diverse sources. Table 2 presents the absolute number of observations in each method choice category as they are derived from the three sources of information. In order to ease the

Table 2. Absolute Number of Quotes Falling into Method Choice Categories as Derived from Three Types of Sources.

	Source Type				
	Group A: Books	Group B: Journals	Group C: Informal	Uncate-gorized	Total
Method Choice Categories					
Positivism/Scientific Paradigm	11	17	5	5	**38**
Technical	23	80	23	11	**137**
Empirical Results	28	56	8	4	**96**
Data-Related Issues	33	47	10	11	**101**
Computation Costs	8	15	6	2	**31**
Difficulty Understanding Method	2	4	2	2	**10**
Subjective Pragmatism	12	25	17	8	**62**
Method/Theory Correspondence	20	49	9	1	**79**
Range of Economic Problems	2	0	1	1	**4**
Underlying Assumptions	4	43	3	2	**52**
Philosophy/Conceptualization	16	18	9	7	**50**
Methodology	20	21	11	3	**55**
Aesthetics	26	18	5	8	**57**
Background/Training	2	5	6	7	**20**
Individuals Involved/Acquaintances	5	1	6	1	**13**
Marketing	6	8	6	3	**23**
Interdisciplinary Communication	0	1	0	0	**1**
Total	**218**	**408**	**127**	**76**	**829**

reading of this table and further information, these three sources of information are defined in the column heading with reference to the primary type of resource relied upon in that grouping. The values in Table 2 are the material upon which the chi-square tests are run.[8]

The first test is on differences in the decision criteria as they are obtained from formal versus informal sources. The formal category of sources combines Group A (books) with Group B (journals). Group C provides the observations for informal sources. The hypothesis undergoing test is that there are no differences among decision criteria frequency with respect to source type. The estimated value of the test statistic is 34.53, and the hypothesis must be rejected at the 0.001 level of significance with 13 degrees of freedom.[9] In other words, the results of this test suggests that the two samples of information divided by formal and informal sources do contain distinct sets of information on method choice, satisfying expectations. This test does not provide evidence which could be used to refute the Knorr-Cetina hypothesis of formal materials polishing out some of the events occurring in the scientific workplace.

The second hypothesis is that there are no differences in the information from the two groups which constitute the formal observations. The briefly confounding result obtained from an estimated chi-square of 42.60 is to reject this hypothesis between books and journals at the 0.001 level of significance with 12 degrees of freedom.[10] In other words, books and journals do *not* provide similar information on method choice. This finding is surprising because it confounds the expectation that all formal sources will yield similar information.

If journals and books are not the same, is the information derived from books alone similar to that coming from informal sources of information? The chi-square test on the specific categories suggests that the information from these two sources of information *are* similar. At the 0.001 level of significance with 10 degrees of freedom, we can not reject the hypothesis that there are no differences among decision criteria with respect to source type.[11] The result suggesting similarity between books and informal sources is further validated in a chi-square test of the observations for these two source types as they fall into the general categories on method choice. Finally, since books more closely resemble informal sources, a final test is established to determine if journals are distinct from this combined group. In this final test, the first sample consists of information derived from journal articles in the history of econometrics (e.g. Group B), and the second sample includes the combined observations obtained from books and from informal sources (e.g. Group A plus Group C.) The estimated value of the chi-square on this final test is 50.71, and so the hypothesis that there is no difference between the two sources of information must be soundly rejected at the 0.001 level of significance with 12 degrees of freedom.[12]

How do all these statistical tests translate into common sense about method choice in the history of econometrics? First, the source of information used by the historians affects the stories they tell about method choice because different sources appear to be related to different aspects of the method choice taxonomy. The second finding is that the type of information on method choice found in books differs in some ways from information found in journals. Although both are polished and published accounts of econometric research, something else is going on in the accounting of factors influencing method choice in these two sources of information besides concerns regarding external scientific approval for publication. Third, books and informal sources appear to

Table 3. Number of Quotes Falling into General Method Choice Category from a Specific Source as a Percentage of Total Number of Quotes Derived from Source Type.

Source Type	Group A: Books %	Group C: Informal %	Group A + Group C %	Group B: Journals %	Uncate- gorized %	Total
Categories						
Science	28	28	28	38	26	**33**
Pragmatism	25	28	26	22	30	**25**
Suitability	12	10	11	23	5	**16**
Metaphysics	28	20	25	14	24	**20**
Sociology	6	14	9	4	14	**7**
Total (100%)	**218**	**127**	**345**	**408**	**76**	**829**

be vehicles from which similar information is communicated at least on the issue of method choice, and the information derived from these two sources diverges from journal testimonials. The remainder of this section is devoted to an evaluation of the specific categorical differences apparent across source types.

Table 3 provides the breakdown of the frequency of observations in the general categories of method choice as they are derived from particular types of sources. This information is presented as the percentage of observations obtained in a particular method choice category from a specific source of information as it relates to the total number of observations for that source. The sources are grouped as follows. The first column is based on information from books. The second column is information from informal sources of information and the third column combines the observations from the first two columns. The fourth column presents the information derived from journals. Finally, there are quotes for which no source of information was available; they are accounted for in the uncategorized column of the table. For example, 28% of the information derived from books concern scientific factors to choose methods, and 38% of the information derived from journals concern scientific factors.

Based on the results of the chi-square tests reported above it was found that there are differences in the type of information obtained depending on the source of information used. Table 3 is presented to help determine in which categories these differences exist. Reading across the rows of the table on percentages allows one to determine the sources to choose if one knows in advance what kind of information on method choice is desired. The following numerical conclusions are based on the sample's density of observations for a particular source type for the general method choice categories. Journals are the source of information in which we find the most frequent reports of scientific decision criteria, with 38% of all the observations in journals related to this topic in the taxonomy. Information on pragmatic reasons to choose methods can be derived from all of the source types, as it is the general category with the most even distribution of information across sources. Journals stand out as the primary source of information for reasons associated with the suitability of a method for the problem under consideration, with 23% of all the information derived from journals falling into this category. Books are the primary source of information for metaphysical aspects of method choice, followed by informal sources of information. There is a definite discrepancy between the combination of information from these two sources in the general category of metaphysics as compared to information on metaphysics from journals. Twenty-five percent of all of the information derived from books and informal sources are concerned with metaphysical issues, whereas only 14% of all the information from journals concerns metaphysics. Books are likely to be the best bet for those interested in the metaphysical aspects of method choice. Finally, of the limited information available on the sociological grouping the numbers suggest that relative to the other categories, the informal sources of information may be the most fruitful place to search for sociological factors in method choice. The differences in percentages informally reinforce Knorr-Cetina's argument that the report of research in the formal journal article tends to highlight the scientific aspects of research activity and tends to polish out sociological aspects. This outcome is important for the methodology of the historian of twentieth century economic thought; to obtain a more holistic vision of twentieth century economics, the historian of thought must rely upon both informal and formal sources of information.

Looking down the columns in Table 3, we can determine the type of method choice information which a particular source is likely to yield. Books are an illuminating source for information on scientific, pragmatic and metaphysical factors but yield less on "suitability" of a method to task and on sociological factors. As with books, informal sources are a fine place to obtain information on scientific, pragmatic and metaphysical concerns, also yielding little on sociological factors or suitability. Finally, in journals scientific factors dominate with 38% of the information from this source relating to scientific concerns (especially technical and empirical results), suitability comes in second, and pragmatism third. As one would expect, journal articles are not the source from which to obtain information on the sociological aspects of method choice.

When the two breakdowns are compared between journals and the combination of information from informal sources and books, journals have more information on scientific aspects in method choice and in suitability, the other two are higher in metaphysics and twice as high in sociology (given the limited numbers on the latter category.)

The concern of this section has been to analyze the observations in the method choice framework for differences depending on the source of information from which the observations were derived. It was found that books and journals present differing information on method choice, and journals and informal sources also present differing information, but informal sources have similarities with book sources. The major findings are that all the general decision criteria in the taxonomy receive coverage from all the sources of information, but the extent of coverage varies. Journals are a strong source for scientific factors and also a good place to find information on suitability, but weak on sociological determinants of method choice. This finding supports Knorr-Cetina's study which found that journal articles tend to polish away context-based information regarding scientific practice. Books differ from journals in that they are a strong source for metaphysical decision criteria.[13] The most likely place to find information on sociological factors is from informal sources of information, although there is a problem in the small number of observations available in this general category because of the historians' limited citations of source material from which the sociological findings were derived.

V. CONCLUSION

In broad terms, the focus of this research has been on decision making in the scientific workplace. More specifically, attention centered on the study of factors influencing method choice in the history of econometrics. One of the primary contributions of this work to the history of economic thought is the development of a method choice taxonomy. This taxonomy was specifically designed as a framework in which to place the decision criteria which have been involved in method choice in the history of econometrics. The major methodological insight from the taxonomy lies in the acknowledgement of a broad range of categories which apply to scientific decision making in the history of econometrics. Although derived from an analysis of the history of econometrics, the specific categories are believed to translate to other matters of choice in science; the general categories are easily extensible and have the added characteristic of flexibility when considering other areas in scientific decision making. The framework is also a convenient foundation for further methodological study.

The method choice taxonomy was applied to consider the relationship between the decision criteria and the source in which the information on method choice was preserved. The importance of this study was based on the hypothesis that formal sources polish away particular types of information regarding scientific practice. The results suggested that although all the categories received coverage by all the source types, different sources emphasize different types of information.

Econometric practice in the twentieth century is an extraordinarily rich subject. The documentation of this history has become the subject of research for a small but growing number of historians of thought during the past decade. In contrast, this paper has concentrated on the nature of the development of this science, in particular the practice of scientific decision making in regard to method choice. It is believed that the contributions of this research on the history of econometrics will be helpful to both historians and econometricians. Historians will benefit by becoming fully cognizant of the general set of factors involved in method choice decisions, and their histories would improve by the recognition of a consistent set of criteria involved in scientific decision making. Furthermore, historians interested in obtaining methodological insights into scientific practice will benefit by applying the taxonomical scheme to questions of a methodological nature. Econometricians will benefit from this work through a deeper understanding of the methodological practices in the history of the field and an explicit understanding of the factors which influence decisions in the scientific workplace. The most fruitful area for continued methodological study using the method choice taxonomy lies in the analysis of the decision making of econometricians from original source material spanning formal, polished and informal accounts of research. It would be particularly helpful to survey econometricians regarding their method choice methodology and to determine what informal sources of information are available for further scrutiny of scientific decision making in econometrics. Interesting comparative studies could also be made of the decision making of theoretical and applied econometricians.

NOTES

1. By adhering to a general definition of "method," over 130 different variations of methods are considered for this work. In Madden (1995), these methods are grouped into seven types, including: programs, models, time series econometrics, estimation techniques, data adjustment techniques, tests/goodness of fit methods, and a

miscellaneous category. Assuming that different method types are chosen for different reasons, these differences are elaborated in (1995) for each of the method types. Although differences in the frequency of citation are noteworthy depending upon the type of method being considered for use, it is not uncommon for any of the method choice criteria (described later in this paper) to be referenced for any of the method types considered. Appendix A provides the list of methods broken down by type which are considered in this project.

2. The following are examples of quotes extracted from the histories of thought and used in this project. A method requirement was that it must be objectively laid out so that the steps can be "reproduced exactly by an individual working according to the direction laid down. The primary requirement in the argument is that each step be tested. The primary requirement in the method evolved is that the operations should be fully explained and reproducible" (Morgan, 1990, p. 58); "technical difficulties" (ibid., p. 50); the "Great Depression was commonly believed to have seen the decline from academic respectability (and perhaps a temporary wane from commercial use) of business barometers" (ibid., p. 67); method choice depended on the provision of "better fit" (Epstein, 1987, p. 21); "method of studies ... is conditioned by the ... characteristics of economic data;" (Christ, p. 31); the Cowles Commission was founded "in the hope that the application of mathematical methods ... would lead to better predictions of stock market behavior" (Epstein, 1987, p. 60); if a method "was not seen as offering a great return in terms of devising development strategies and other very ambitious economic policies" the tool could be rejected (ibid., p. 149); H. L. Moore's desire "to refute Socialist notions of labor exploitation" influenced his choice of technique (ibid., p. 13); "expositional simplicity" (ibid., p. 24); "Frisch was one of the pre-eminent econometricians of the 1930's and consequentially confluence analysis was well known" (Hendry and Morgan, p. 44); a method was not popular because it was published in "an extremely obscure appendix" "tucked away in a book devoted to tariff problems" (Epstein, 1987, p. 29; Morgan, p. 188); and it was noted that "econometricians in that era were already under pressure from the journals to avoid complicated mathematics" (Epstein, 1987, p. 24). Many more quotes were extracted; a review of chapter 4 in Madden (1995) provides method choice category-specific examples of these quotes. All of the quotes used in the study are available upon request. In the statistical analysis in Section IV only quotes relevant to specifically named econometric researchers were used.

3. In following this iterative approach to define the categories of the method choice taxonomy, the definitions themselves were written to delineate the placement of quotes into appropriate categories as clearly as possible. Each reason to choose a method was placed in only one category. In some categories, the final placement of quotations relied upon a judgment call as to the dividing line between related categories; in these cases, the definition of the category was written to allow explicit placement of the quotes in the appropriate category. For instance, when considering sociological aspects to method choice, the description of the category "background/training" includes the influence of a mentor on a mentee, which obviously overlaps with the "specific individuals involved/ aquaintances" category. When a historian explicitly mentioned a mentor/mentee relationship as a reason for method choice, that reason was classified under background/ training and was not also counted in "specific individuals involved/acquaintances." The categories "method/theory correspondence" and "underlying philosophy/conceptualization" also have fuzzy boundary problems. Whereas method/theory correspondence

consists of method choice reasons explicitly relating theory to method, when economic theory itself (and thus the corresponding method) is rejected, the quote is categorized in "underlying philosophy/conceptualization." Categorical definitions provided in this paper are a summary of more detailed definitions in Madden (1995).

4. The idea of using a statistical approach to study methodological issues in econometric history occurred to me as a natural result of the methodology I used to analyze method choice. After I completed the statistical sections of my work in method choice my advisor, Vincent Tarascio, pointed out the parallels between my statistical approach and some of George Stigler's work in the history of economic thought. See the bibliography for reference to one of Stigler's papers.

5. The 1,181 observations in the data set used to develop the method choice taxonomy includes two different sub-sets of information: (1) an individuals data set with 829 observations on individual economists; and (2) a non-individuals data set with 352 observations. The individuals data set is constituted of specifically named decision makers: designated economists, econometricians or scientists in the history of economic thought involved in research related to economics and a few well-defined small groups whose members are readily known. The non-individuals data set include references to decisions made by broad groups of decision makers (e.g. "economists," "econometricians"), the historians' decisions, and the undefined decision makers.

6. Note that the econometricians working in the early twentieth century did theoretical and applied econometrics. As pointed out by an anonymous referee "during the period her sources cover, almost all the econometricians were 'applied economists' who very explicitly saw themselves as people figuring out ways to use statistical methods to solve pressing problems of economic theory and policy. They were not like the econometric theorists of today, who prove theorems for the joy of it. I think they felt a burden at the time of proving the utility of statistical work to an ambivalent profession." Thus, the econometricians in this study, although developers of technique, were also making choices in technique and are an appropriate group to use in this study. A few examples of men in this study who were defining and applying econometrics include Henry L. Moore, Henry Schultz, Wesley C. Mitchell, Warren M. Persons, Jan Tinbergen, and Lawrence Klein. There are many more.

7. The complexity of the relationship between method choice and computability has been remarked upon by Greg Madden and by an anonymous referee.

8. I followed a chi-squared procedure from Hubert M. Blalock, Jr.'s *Social Statistics*, Revised Second Edition. The chi-square approximation used in this paper is given by

$$\chi^2 = \Sigma \frac{(f_o - f_e)^2}{fe}$$

where f_o refers to the observed frequency and f_e refers to the expected frequency for each cell.

9. The degrees of freedom drop from 17 to 13 because of a sample size less than 6 observations in some of the categories. Specifically, "positivism/scientific paradigm" was combined with "methodology", "range" was combined with "method/theory correspondence", and "interdisciplinary communication" was combined with "marketing".

10. In this instance, due to small sample size the observations for "range" were combined with "method/theory correspondence" and the entire sociological section was collapsed to a single sociological category.

11. Due to small frequencies of observation, the numbers for "positivism" were combined with "methodology", the specific categories in suitability were collapsed to one category, and the specific categories in sociology were collapsed to one.

12. The observations in the sociological section were collapsed into a single method choice category for this test and the observations on "range" were combined with those in "method/theory correspondence".

13. In method choice, there may exist an informal disciplinary hierarchy of method choice reasons, flowing roughly in accordance with the method choice taxonomy (e.g. science first, sociology last, if at all). An anonymous referee pointed out that since space constraints dominate in journal articles, it seems likely that the reasons at the top of the disciplinary method choice hierarchy would be more concentrated in journal articles than in books. A concentration measure can be tabulated for books and journals. The top five categories reported in journals are technical reasons, empirical results, method/ theory correspondence, data-related issues and underlying assumptions. The citation of these reasons in journal articles as a percent of all the mentioned reasons in journal articles is 0.67; in contrast this number is 0.495 in books. The top five categories reported in books are data-related issues, empirical results, aesthetics, technical reasons and methodology. The citation of these reasons in books as a percent of all the mentioned reasons in books is 0.6; in contrast this number is 0.54 for journal articles. Thus, journal articles do appear to have a higher concentration ratio in five of the categories which may rank most highly in method choice.

ACKNOWLEDGMENTS

This chapter is the result of research for my doctoral dissertation at the University of North Carolina at Chapel Hill. I would like to thank my advisor, Vincent Tarascio, two anonymous referees, and Greg Madden for feedback regarding this work.

REFERENCES

Blalock, H. M., Jr. (1979). *Social Statistics*, revised 2nd ed. New York: McGraw-Hill.

Broster, E. J. (1937). A Simple Method of Deriving Demand Curves. *Journal of the Royal Statistical Society, 100*, 625–641.

Christ, C. F. (1952). *Economic Theory and Measurement: A Twenty Year Research Report, 1932–1952*. Baltimore, Md: Waverly Press.

Madden, K. K. (1995). Method Choice in the History of Econometrics: A Study of Decision Criteria Involved in the Selection of Econometric Techniques as Recorded by Historians of Econometric Thought. Ph.D. diss., University of North Carolina at Chapel Hill.

de Marchi, N., & Gilbert, C. L. (Eds) (1989). *History and Methodology of Econometrics*. Oxford: Clarendon Press.

Epstein, R. J. (1987). *A History of Econometrics*. Amsterdam: North-Holland.

Epstein, R. J. (1989). The Fall of OLS in Structural Estimation. In: N. de Marchi & C. L. Gilbert (Eds), *History and Methodology of Econometrics* (pp. 94–107). Oxford: Clarendon Press.

Gilbert, C. L. (1989). LSE and the British Approach to Time Series Econometrics. In: N. de Marchi & C. L. Gilbert (Eds), *History and Methodology of Econometrics* (pp. 108–130). Oxford: Clarendon Press.

Haavelmo, T. (1943). Statistical Testing of Business-Cycle Theories. *Review of Economic Statistics, 25*, 13–18.

Haavelmo, T. (1944). The Probability Approach in Econometrics. Supplement to *Econometrica, 12.*

Hendry, D., & Morgan, M. (1989). A Re-Analysis of Confluence Analysis. In: N. de Marchi & C. L. Gilbert (Eds), *History and Methodology of Econometrics* (pp. 108–130). Oxford: Clarendon Press.

Knorr-Cetina, K. D. (1981). *The Manufacture of Knowledge: An Essay on the Constructivist and Contextual Nature of Science.* Oxford: Pergamon Press.

Koopmans, T. (Ed.) (1950). *Cowles Commission Monograph 10: Statistical Inference in Dynamic Economic Models.* New York: Wiley.

Kuhn, T. S. (1970). *The Structure of Scientific Revolutions.* Chicago: University of Chicago Press.

Mann, H. B., & Wald, A. (1943). On the Statistical Treatment of Linear Stochastic Difference Equations. *Econometrica, 11*, 173–220.

Mitchell, W. C. (1913). *Business Cycles and their Causes.* Berkeley: California University Memoirs.

Mitchell, W. C. (1927). *Business Cycles: The Problem and its Setting.* New York: National Bureau of Economic Research.

Moore, H. L. (1914). *Economic Cycles – Their Law and Cause.* New York: Macmillan.

Moore, H. L. (1917). *Forecasting the Yield and the Price of Cotton.* New York: Macmillan.

Morgan, M. S. (1990). *The History of Econometric Ideas.* Cambridge: Cambridge University Press.

Persons, W. M. (1919). Indices of Business Conditions. *Review of Economic Statistics, 1*, 5–110.

Persons, W. M. (1919a). An Index of General Business Conditions. *Review of Economic Statistics, 1*, 111–205.

Pickering, A. (1995). *The Mangle of Practice: Time, Agency, and Science.* Chicago: The University of Chicago Press.

Stigler, G. J. (1969). Does Economics Have a Useful Past? *History of Political Economy, 1*, 217–230.

Tarascio, V., & Caldwell, B. (1979). Theory Choice in Economics: Philosophy and Practice. *Journal of Economic Issues, 13*, 983–1006.

White, H. (1987). The Value of Narrativity in the Representation of Reality. In: W. J. T. Mitchell (Ed.), *On Narrative* (pp. 1–24). Chicago: University of Chicago.

APPENDIX A

A Method Typology

Programs

econometrics
macrodynamic model building and estimation
..iterative approach between theory and empirical work
probability approach
..probability as a foundation for confluence analysis
structural estimation
business cycle indices, including:
..Harvard A-B-C curves
..Berlin Institute system of indices
statistical economics of Mitchell
Burns/Mitchell (NBER) approach, including
..specific and reference cycles
confluence analysis
Bayesian approach

Models

single equation models, including:
..function fitting by inspection
..regression of p on q
..regression of q on p
..dynamic single equation models
..dynamic single equation model with integral
cobweb model
two-equation supply and demand models
reduced form (Tinbergen's derivation)
linear difference regression equation
linear differential equations
rocking horse model
sequence analysis
macrodynamic models (difference equation models)
Tinbergen's macromodel for the League of Nations
Tinbergen's final equation method (difference equation)
Tinbergen's 'mongrel' equations (modified reduced form
estimating curves separately)
zero restrictions on a system of structural equations

Models, continued

structural models
reduced form of complete equation system
structural form of complete equation system
simultaneous equations models
index numbers in structural estimation
recursive model
Klein/Cowles Commission structural models
identifying restrictions in macroeconomic models
identifying restrictions on solved reduced form of structural models
least squares reduced form of structural models (no identifying restrictions)
solved reduced form of structural models with identifying restrictions
distributed lag consumption function model
Klein's complex consumption functions
naive consumption function model
textbook consumption function model

Time Series Econometrics

frequency domain approach, including:
..fourier analysis
..harmonic analysis
..periodogram analysis
..moving method
..Davis' modification of periodogram
time series decomposition analysis
..direct trendfitting by means of curve fitting
..trendfitting by means of moving averages
correlogram analysis
counting of peaks and troughs
variate difference method
closed loop control system
autoregressive moving average model
vector autoregression

Estimation Techniques

correlation analysis, including:
..time series correlations on 2 positively serially correlated series
..time series correlations on 2 positively serially correlated series with
positively serially correlated differences

Estimation Techniques, continued

..time series correlations on oscillatory series
graphic methods of correlation
Tinbergen's graphic stacking technique
regression analysis
orthogonal regression
orthogonal regression on regression residuals
Wright's 'successive slopes' method
Pigou's method to obtain demand elasticity
Leontief's data splitting method to derive elasticities
Broster's method to derive demand curves
ordinary least squares (OLS) analysis, including:
..OLS on single equation models
..OLS on single stationary linear stochastic difference equation
..OLS on systems of equations
..OLS on structural models
..OLS on simultaneous equations models
..OLS on reduced form
..OLS on structural form
..OLS on autoregressive stochastic difference equation models
..OLS on dynamic models
..OLS on recursive model
..OLS on exactly identified two equation model with first order autocorrelated
errors, serial correlation coefficient of one
..OLS on integrated processes
indirect least squares (ILS)
maximum likelihood estimation methods (MLE), including:
..full information maximum likelihood (FIML/MLE)
..MLE on structural models
..MLE on simultaneous equations models
..MLE on structural form of complete equation system
..MLE on reduced form of complete structural equation system
..FIML on exactly identified two equation model with first order autocorrelated
errors, serial correlation coefficient of one
..limited information maximum likelihood (LIML)
..LIML on structural models
Tukey's alternative structural estimation procedure (to FIML)
two-stage least squares (2SLS)
fix point method to estimate systems, based on 2SLS

Estimation Techniques, continued

path analysis
Wright's instrumental variables method
Tinbergen's final equation method (difference equation), including:
..from initial values, extrapolate further values of final equation
..mathematically advanced: solve final equation for harmonic terms
multi-model estimates method (estimates based on a priori degrees of
confidence, functions of estimates of variable coefficients)
autoregressive least squares (ALS)

Data Adjustment Techniques

percent difference between actual curve and "normal" curve (adjusted growth
curve)
percent deviations from sample means
trend ratios
differences from nine year moving average
first difference transformation
link relatives
per capita data
time trend in equation

Tests

standard errors
bunch maps
out of sample forecasting tests of structural models
within sample significance tests of structural models
classical testing methods
asymptotic significance tests
confidence tests (e.g. t ratios)
confidence intervals
von Neumann ratio test for serially correlated residuals on structural models
Anderson's test for statistical validity of overidentifying restrictions
use of panel data as foundation for making exogeneity decisions in structural
models

Miscellaneous

averaging data
mean deviations
ranking data
statistics
summary statistics (means and variances)
mathematical methods (graphs, equations, calculus, algebra)
"method of ceteris paribus"
close observation of relevant case material
formal logic
interview method
judgment

APPENDIX B

Decision Period	Percent of 829 Observations Falling in Decision Period
1860–69	0.36%
1870–79	0.97%
1880–89	0.12%
1890–99	0.12%
1900–1909	0.00%
1910–1919	4.58%
1920–1929	13.27%
1930–1939	21.47%
1940–1949	34.98%
1950–1959	11.22%
1960–1969	3.62%
1970–1979	0.84%
1980–1989	2.05%
No Date	6.39%

UNPACKING "ADAM SMITH: CRITICAL THEORIST?"

Spencer J. Pack

ABSTRACT

This paper responds to Keith Tribe's provocative Journal of Economic Literature *article, "Adam Smith: Critical Theorist?" There Tribe argued that most people most of the time grossly misread Smith, due, among other things, to their quite inadequate appreciation of Smith's linguistic, social, moral, and theological context. Against Tribe, the paper argues that Smith can profitably be read as* both *an eighteenth-century moralist and a twenty-first century critic. Smith can be a source of inspiration, wisdom and profundity for contemporary economists. Moreover, Smith can be successfully employed by modern economists to change, deepen, and broaden contemporary economic theory.*

I. INTRODUCTION

This article clarifies and draws out the implications of Keith Tribe's provocative piece "Adam Smith Critical Theorist?" (1999). Part II recalls Professor Tribe's sharp criticisms of most people who attempt to read Smith. Essentially, according to Tribe, these readers go wrong because it is so intrinsically difficult for contemporary readers to understand Smith. Part III unpacks Tribe's presentation of all the supposed difficulties and dangers involved in attempting to read Smith. Part IV then argues that in effect Tribe's timid, overly cautious, fearful approach overestimates the costs and underestimates the benefits of

Research in the History of Economic Thought and Methodology, Volume 19A,
pages 33–46.

reading Smith. Tribe's faulty accounting unduly discourages contemporary economists from studying Smith, leads to an under-investment in Smithian scholarship, and cuts contemporary economists off from what Smith has to offer. Therefore, Tribe's fundamental approach to Smith's work needs to be rejected since it unduly discourages economists from directly studying and developing Smith's thought. Part V concludes that contemporary readers will find that Smith was deeply critical of his eighteenth century society. They will also discover that Smith can be used to help critique the problems and ills of our own society; hence, that Smith can indeed be viewed to be what Tribe calls a "critical theorist". They will see that Smith's work can be used to broaden and deepen contemporary economic theory; by directly encountering and studying Smith, contemporary economists will be able to further promote and develop the rich legacy Smith has bequeathed to us.

II. PROFESSOR TRIBE'S CRITICISMS

In a provocative article in the *Journal of Economic Literature*, Keith Tribe has sharp words for most people who attempt to read Adam Smith. He criticizes "senior economists who practice as amateur historians" such as George Stigler (1975), whom he says conflates Smith's position with that of Mandeville (622). In addition, Joan Robinson (1964) misunderstood Adam Smith as trying to abolish the moral problem, thus also demonstrating her inadequate understanding of the relationship between Smith and Mandeville (620); recently Kenneth Lux (1990) made the same mistake (620, fn. 44). Edward West (1996) naively attributes "mistakes" or "confusion" to Smith's work (615, fn. 30). Unlike the putative skilled historian, West does not realize that Smith's "significance and meaning is not immediately accessible to us" (615); thus, West unfortunately applies "an inappropriate conceptual grid" (616). Patricia Werhane (1991) has a "kind of sloppy argument" and she displays a "lack of clarity and coherence" (628). Her work and my book on Smith (Pack 1991) can be "lightly dismissed" (629) because we give insufficient attention to the historical specificity of Smith's "project". Jerry Muller, a historian, and I "have credentials the mainstream economist might consider suspect" (610).[1] Muller (1993) is criticized for using Smith's *Theory of Moral Sentiments* to construct Smith's moral philosophy and the later sections of Smith's *Wealth of Nations* to deal with institutions needed in modern societies; he fails to realize "the dangers of creating a unitary construct out of material drawn from diverse sources" (629). The diverse sources are, of course, two completely different books (or "texts") written by the same author, Adam Smith. Michael Shapiro's work (1993) is so bad that "one would hope that there is nothing worse out there" (619, fn 41). Books by Peter Minowitz (1993) and Athol Fitzgibbons (1995) are flawed because of their theological naivete (625); and, Deborah Redman (1997) "recycles" the "old idea" that "the new science of political economy was avowedly secular" (625). Thus, for Tribe, most people most of the time grossly misread Adam Smith. That is why, according to Tribe, the "history of attenuation, misreading and misunderstanding, of which *Das Adam Smith Problem* is just one part, is the real historical Smith" (630, italics in original). Indeed, instructs Tribe, the "actual reception process" of Smith's works is really the story of "so many failures of comprehension" (ibid). Why is this? What is going on? The answer for Tribe is actually quite simple: reading Adam Smith is so extremely difficult. For Tribe, it can also be rather dangerous precisely because Smith is so hard to comprehend. Now, there are those, such as the eminent John Kenneth Galbraith, who consider Smith to be the very best writer among English-speaking economists (1971); Smith is generally held to be an extremely clear writer. Therefore, it may be of keen interest to look into Tribe's surprising thesis in more detail.

III. UNPACKING PROFESSOR TRIBE'S "DISCOURSE"

Tribe begins his article with a quote from the venerable Thomas Hobbes. Part of it says, "It must be *extreme hard* to find out the opinions and meanings of those men that are gone from us long ago, and have left no other signification thereof but their books" (Hobbes, *The Elements of Law, Natural and Politic* I.13.8, quoted in Tribe, 609, emphasis added by me).[2] Tribe refers back to this quote in the main text when he writes that ". . . following the rubric laid down by Hobbes at the beginning of this essay, if we are to deepen our understanding of the work of Adam Smith, we need to reexamine well-worn assumptions concerning his work" (616). For Tribe, following contemporary trends in text-reading, this means that we have to really pay attention to language. Thus, instructs Tribe, ". . . the protocols governing the writing of intellectual history" (of which the history of economic thought is one part) currently dictate that we must pay "close attention to language, its use, and context" (617). It turns out that this is not so easy, since, following the latter Wittgenstein , "the meaning of a word lies in its use". Hence, we must "attend to the occasions upon which particular words are used" (ibid). However, "this in turn requires that we are familiar with the relevant linguistic field". Yet, this means that we need to understand not only "explicit statements", but also "implicit omissions" (ibid). Moreover, this cannot be done without "a detailed understanding of the appropriate linguistic context" (ibid).

Thus, Tribe writes that "if we are to read Smith" (and at this point the intelligent economist may be wondering if she *really* does want to read Smith) then "we need to pay especial attention to what he did and did not say; we should respect the integrity of the language he employs" (623). Tribe points out that people in the history of political *thought* have transformed that field into the history of political *language* (618, emphasis by Tribe). Presumably, for Tribe, the history of economic thought should also be morphed into the history of economic language. Thus, sophisticated theoreticians realize that "our understanding of *what* was being said or written depended on *how* or *where* it was being said" (ibid). Hence, we need to derive meaning "through an analysis of language structure" (ibid). All of this would be much easier to do if economists understood the work of Saussure in linguistics and the philosophy of Wittgenstein (the latter, not the earlier Wittgenstein). Therefore, according to Tribe, in reading Smith, "neglecting the contemporary linguistic context" is dangerous (623); indeed, this is what got Stigler in so much trouble in his above-noted flawed attempt to read Smith.

Moreover, it might also be useful to understand "the Stoic tradition within which Smith worked" (625). And do not forget theology, "a matter of great importance to the early development of political economy" (ibid). Readers of Smith need to understand "the role of natural theology in the shaping of Divine Providence" (627). The fact that theology "has generally been neglected by historians is once again evidence of the *dangers* of imposing a modern secular perspective upon writers of the past" (625, emphasis added). Clearly, reading Smith is not only difficult and demanding; it is also dangerous.

It might also not hurt if readers understood the work of Jacques Derrida or Michel Foucault (619). Furthermore, according to Tribe, "Smith's argument has to be viewed in the context of an eighteenth-century debate on commerce and civilization" (620). According to Tribe, it is unfortunate that the conventional approach to studying the history of economic thought assumes "we already know what an 'economic text' is" and it evaluates the texts solely "in terms of modern problems and issues" (616). Tribe feels this approach ought to be rejected.

Instead, for Smith, we need "careful attention to the context and reception of his writings" (626). Unfortunately, another problem is that "we still know all too little about the details of the subsequent reception process" to Smith's writings (615). Things are indeed difficult for Professor Tribe because it is "only after we have gained a better understanding of Smith's 'project' (evidence for which is at best fragmentary) that we can then move on to identify with any accuracy the manner in which this was communicated through the work of succeeding generations" (630). So, according to Tribe, we must first understand Smith's project; then we need to understand how it was read by his readers. Note that, basically, for Tribe, history becomes the story of book readers. Real economic, political and social history tends to be completely ignored by Tribe. Thus, by Tribe's reckoning, it is "the history of editions and rereadings that separates our appreciation of Adam Smith from that of his contemporaries" (615, fn. 29). Yet, for Tribe, all is not bleak. In the perhaps not-too-distant future, we will be able to "reconstruct Smith's arguments on the basis of his language and its contexts" (630). This will "indeed provide us with a new Smith and a fresh understanding of his analysis of commercial society; and in the process he will certainly re-emerge as a critic . . . of features of a commercial society" (ibid). Hence, Tribe concludes that we do not yet know that much about Smith. Yet, with care and research, we will in the future. Only then will Smith appear as a critic of some particular features of a commercial society.

Tribe approves of relatively few readers of Smith. He does highly recommend the work of Brown (1994), Hont and Ignatieff (1983a) , Hundert (1994), and Winch (1978; 1996). This is not the place to present detailed criticisms of their work.[3] All of these authors are indeed careful, diligent writers. They are all also extremely cautious and hesitant to relate Smith's works and ideas to contemporary concerns and issues. This caution receives Tribe's uncritical approbation.[4] On the other side of the field, Tribe cites the work of Muller (1993), Werhane (1991) and myself (Pack, 1991). We are particularly concerned to try to relate Smith's work to contemporary concerns. The titles or subtitles of our books give us away: *Adam Smith in His Time and Ours*; *Adam Smith and his Legacy for Modern Capitalism*; *Adam Smith's Critique of the Free Market Economy*. Tribe charges us with foolishly attempting to turn Smith into a "critical theorist";[5] that is, a person who can be a "modern critic". For that we are chastised by Tribe, since "we learn from Smith not by converting him into a twentieth-century critic, but by understanding him rather better as an eighteenth-century moralist" (629). Hence, the answer to Tribe's question which titles his article, "Adam Smith: Critical Theorist?" is an unequivocal no.

IV. SMITH AND CONTEMPORARY ECONOMIC READERS

Tribe's approach to reading Smith is too cautious, timid, and fearful. It needs to be rejected for several reasons.

Obviously, as the recent work in rhetoric and textural criticism emphasizes,[6] there is no one "right" way to read Adam Smith. Adam Smith is not necessarily *either* an eighteenth-century moralist *or* a twentieth-century critic. He can be both.

Moreover, as Tribe himself writes more than once, Smith has been viewed "as a *prophet* of economic liberalism" (613, emphasis added); "he is the *prophet* of what we call modern capitalism" (619, emphasis added). Naturally, people, including many politicians, policy advisers, civil servants, and popularizers, will want to know what the prophet would have to say about contemporary society. Of course, as Tribe emphasizes, Smith has been dead quite a few years, and hence it is impossible to say with certainty exactly what he would say now. Yet, economists as public intellectuals, do have a responsibility to try to set the record straight. We need to try to inform the public about what Smith's views would likely be on contemporary society, so that Smith's legacy is not captured by the vested interests (Noonan 1990, noted in Galbraith 1992: 98 ; Lerner 1937: X) or even "madmen in authority, who hear voices in the air" (Keynes 1964: 383).

Yet, more than this, Smith himself can be used to as a source of inspiration, wisdom and profundity for contemporary economists. We can indeed benefit by having the intellectual tool and ability to ask ourselves, "well, what would Smith likely think on this issue or problem?" Smith viewed himself as a "critical theorist". He wrote in a letter of "the very violent attack I had made upon the whole commercial system of Great Britain" (1987: 251, Letter No. 208). Smith can be used to criticize contemporary society and public policy partly because of Smith's own deeply developed sense of history (Smith, 1978), and because history itself has not changed that much (Coats 1994: 147–157);[7] thus his ideas and approach to life are still deeply relevant. As John Kenneth Galbraith has pointed out, if Smith "was a prophet of the new, he was even more an enemy of the old. Nor can one read *Wealth of Nations* without sensing his joy in afflicting the comfortable, causing distress to those who professed the convenient and traditional ideas and policies of his time. There was much in Smith that prescribed sensibly for the new world of which he stood on the edge; his larger contribution was in destroying the old world and thus leading the way for what was to come." (1987: 59–60)[8] Thus continues Galbraith, *The Wealth of Nations* "with the Bible and Marx's *Capital*, [is] one of the three books that the questionably literate feel they are allowed to cite without having read. Especially in Smith's case this is a grave loss" (ibid 62).

This is not to deny that Smith's work is completely without ambiguity or difficulty. It is to say that Smith's work can be and is a source of inspiration for contemporary critical theorists such as Galbraith and others; and that Smith's work can be a valuable tool for contemporary economists. Moreover, it needs to be emphasized that Tribe's overly cautious approach unnecessarily exaggerates the complexities of reading Smith. For example, briefly consider Professor West's piece on the effects of the division of labor upon the worker (1996). Recall that this effort was denigrated by Tribe for applying "an inappropriate conceptual grid" and that West did not realize that Smith's "significance and meaning is not immediately accessible to us". Nonetheless, in defense of West, if one looks at the *very* first sentence of *The Wealth of Nations*, one will read that "The greatest improvement in the productive powers of labour, and the greater part of the skill, dexterity, and judgment with which it is any where directed, or applied, seem to have been the effects of the division of labor" (I. i.1). Yet, hundreds of pages later the patient reader will also read that "The man whose whole life is spent in performing a few simple operations, of which the effects too are, perhaps, always the same, or very nearly the same, has no occasion to exert his understanding, or to exercise his invention in finding out expedients for removing difficulties which never occur. He naturally loses, therefore, the habit of such exertion, and generally becomes as stupid and ignorant as it is possible for a human creature to become." (V.i.f.50)

One does *not* have to be an expert in Foucault, Derrida, or the latter Wittgenstein[9] to realize that there is some tension and ambiguity between these two statements. Moreover, as West himself points out, the public policy implications of understanding and interpreting this "text" can be quite crucial: "At a time when several Eastern European countries are seeking to establish the market system, at least some of their intellectuals may have a desire to scrutinize the origins of Western support for it" (1996: 83). Hence, West is asking the right questions in that they are genuine, topical, and important.

Tribe's approach suggests that the costs of reading Smith are quite high (since it is so difficult and we need to do so much background work to understand him) and the benefits (in the absence of these phenomenal start-up costs) are so low that it seems doubtful whether the rational economist should ever read Smith.[10] As with Keynes where a lack of animal spirits will lead to a rush to liquidity and a lack of real investment, so too, adherence to Tribe's cautious, timid, fearful accounting will lead to a lack of real investment in studying Smith.

Tribe's arguments are particularly surprising given the general clarity of Smith's writing style. Currently, philosophers are rediscovering that Smith is able to address their contemporary concerns and issues. So, for example, Charles Griswold argues that Smith's conception of the virtues, and what it means to lead a virtuous life, is an overlooked, misunderstood resource which may be used to justify and defend the ideals of Enlightenment thought (1999). Samuel Fleischacker (1999) perceptively points out the similarities between the thought of Smith and Immanuel Kant; he argues that they offer us a conception of freedom which focuses upon the freedom to make judgements. This freedom to make informed judgements may be viewed as a more sophisticated, subtle version of the Friedmans' emphasis on "the freedom to choose". Economists are also discovering the applicability of Smith's work to contemporary concerns and issues, such as public choice theory (West, 1990) and game theory (Ortmann, 1999; Ortmann & Meardon, 1995, 1996). Yet, it is not merely an issue of using Smith to address contemporary issues and concerns. Smith is also being used to vigorously and profoundly change and broaden these very perspectives. Hence, Tribe's dictate that "a continuity is established between modern economics and past writings in which only those elements of the latter that can be brought into relation with modern issues and concepts are considered" (616) is too simplistic. Here, for example, Tribe's complete overlooking of the extensive, remarkable work (or "project") of Jerry Evensky (1987, 1989, 1992, 1993a, 1993b, 1994a, 1994b, 1998, Evensky & Malloy, 1994) and Jeffrey Young (1986, 1990, 1992, 1995, 1997) is particularly surprising and regrettable. Evensky and Young are two of the economists who in recent years have subjected Smith to the most searching analyses. Young's articles and book, *Economics as a Moral Science, The Political Economy of Adam Smith*, and Evensky's various articles and edited work, *Adam Smith and the Philosophy of Law and Economics*, together provide a vivid demonstration of how the moral, ethical, religious, legal, and philosophical concerns of Adam Smith can also be used to help broaden, enrich and invigorate contemporary economic theories and research programs. So, for example, Evensky works out "the insights and policy implications to be taken from this analysis of the role of law in Smith's moral philosophy" (1994a: 216) while being keenly aware of the role of economics in Smith's system. Young develops the "rich heritage in Smith which intertwines economics and ethics and which has gone untapped" (1997: 4, fn. 2). Evensky and Young are consciously using Smith to change, deepen, and broaden contemporary economic theory.

Thus, contemporary economists can and are bringing our knowledge of current techniques and concerns to read Smith in new ways. Yet, we can and also are using our knowledge of Smith to inform, deepen and change our contemporary techniques and concerns. What we know and how we are trained as economists necessarily helps to shape our reading of Smith; but, as the recent work of Evensky and Young graphically illustrate, our readings of Smith can in turn change our current concerns and approaches to economics. There can be a dialectical interrelationship between our readings of Smith and present economic thought. As economists, we may read into the past based upon our knowledge and being in the present. Yet, we can also use our knowledge of the past and Smith's work to change our understanding of the present and our own work as practicing, contemporary economists. Thus, an intimate knowledge of the past in general (Walker, 1999), and Smith in particular, can strengthen and transform our knowledge of both the past and the present, and has the potential to help economists nudge society into a better future.

V. READING (NOT REREADING) ADAM SMITH

Tribe is helpful in pointing out that the Six Volume *Glasgow Edition of the Works and Correspondence of Adam Smith* (1976–1983) provides an opportunity to view all of Smith's "writings together – his books, correspondence, surviving lectures, and essays" (615). Yet, Tribe then goes on to write that "this in turn provides a new context for the *rereading* of *The Wealth of Nations*" and to query "How then should we go about the task of *rereading* these writings?" (615, emphases added) Here Tribe is a bit misleading – or over-optimistic. Most readers of the *Journal of Economic Literature*, where his article appeared, are economists; hence, most have probably not read Smith. Or, at least they have not read much Smith, and most likely not in graduate school (Colander & Klamer, 1987). Hence, for most of economists, it is not a question of rereading Smith; it is a question of reading Smith.[11]

Smith's work provides a rich source of information, insight, theory, and inspiration for contemporary economists. Cut off from the past, and particularly from the wisdom of Smith himself, contemporary economists are liable to dig shallowly in the knowledge of the world, their work liable to be easily toppled over by the next intellectual fad or wind of change. Economists should not be frightened by the warnings of Tribe over the supposed dangers or putative enormous difficulties of studying and using Smith. As Tribe insists, Smith was indeed an eighteenth century moralist; but he was and is more than that. When contemporary economists study Smith they will also find a "critique of the free market economy" (Pack, 1991). They will see that "an awareness of Smith's work in the context of his time" will enable them to appreciate "its timelessness and its timeliness" (Muller: 205). They will see that Adam Smith has a "legacy for modern capitalism" (Werhane). And they will be in a position to further develop and enhance that legacy. They will find someone who was a comprehensive critic of his eighteenth century society. They will also find a powerful thinker who, with care, can also be used to help address and critique the problems of the twenty-first century.

NOTES

1. The cause of the suspicion of my credentials is not clear.
2. The identical quote is used by Hundert (1994: viii) to begin his book.

3. Briefly, the following criticisms of their work may be noted. The philosopher Charles Griswold argues that for Brown, even by her own relatively uncharitable reading, which denies authorial intention, there appears to be an underlying Stoic unity to Smith's thought (1995: 30; 1999: 27–28).The political theorist David McNally argues that the work of Hont and Ignatieff (1983b) flattens the rich complexity of Smith's thought by trying to straightjacket it into either a civic humanist or natural jurisprudence tradition (1988: 291, fn.55). The Marxist historian E. P. Thompson criticizes Hont and Ignatieff (1983b) for an overly pretentious writing style, not understanding how economic markets actually work, ignorance of the eighteenth-century newspaper and pamphlet literature, and inadequate understanding of contemporary economic theory, particularly the work of A. K. Sen (1993, Chapter V). The economist Rashid, in a review of Teichgraeber (1986) complains that Winch (1978) fails to demonstrate that he indeed has "a sure grasp of Smith's contributions to economics" (1989: 555).

4. Two of these works are in the "Ideas in Context" series edited by Quentin Skinner: Hundert (1994) is number 31; Winch (1996) is number 39. Perhaps not so coincidentally, Keith Tribe's own book, (number 33) *German Economic Thought from the Enlightenment to the Social Market* is also in the same series.

5. It is not exactly clear what Tribe means by the term critical theorist. There is a school of humanist, cultural Marxists called critical theorists (or the Frankfurt School). Tribe does not refer to the work of such well known Marxist critical theorists as Herbert Marcuse, Theodor Adorno, Walter Benjamin, or Max Horkheimer. Hence, presumably Tribe is not accusing us of attempting to turn Smith into this sort of critical theorist. On this school of thought see Jay (1973). However, the prominent Austrian economist Murray Rothbard (who Tribe also does not mention) does indeed criticize Smith for his critical, radical influence on Marx (1995: Chapter 16; see also Pack, 1997, 1998).

6. For example, Brown writes that "language has a kind of fecundity with a potential proliferation of different readings" (1994: 3); hence, there can be more than one reading of a text. As noted above, Tribe employs Brown's work to criticize Muller for "creating a unitary construct" out of Smith's two published books. This severe use of Brown's stimulating reading seems unjustified. For example, even Winch uses *The Theory of Moral Sentiments* to shed light on *The Wealth of Nations* (1996: 95–96).

7. See also the debate in political theory on the relevance of ancient Greek thought to contemporary society, and whether Greek political theory has become hopelessly anachronistic due to historical changes in Holmes (1979) and Nichols (1979). Surprisingly, Tribe's extreme historicism has the paradoxical effect of devaluing the study of history. Why is this? For Tribe, it is due to dramatic changes, that it becomes too difficult and dangerous to learn and apply the lessons of the past to the present. Actually, economic theorists are frequently guilty of the opposite extreme, ahistoricism, or assuming little or no change throughout the course of history. The ahistorical approach facilitates the relatively uncritical application of contemporary economic theories and techniques to an essentially unchanging past. See, for example, the harsh (though not always accurate) complaints of the philosopher Meikle in dealing with the ancient Greek economy and Aristotle's economic thought (Meikle, 1995). There may be an unholy combination of radical historicism when considering the potential contributions of past thinkers to understanding the present, with an ahistorical approach to using the work of present theorists to study the past. It is this combination which is most conducive to the undue aggrandizement (or hubris) of contemporary economic

science. It implies that past thinkers can teach us very little; but, that current thinkers can adroitly understand the past.

8. This strong "joy in afflicting the comfortable, causing distress to those who professed the convenient and traditional ideas and policies of his time" aptly characterizes John Kenneth Galbraith's own work. In this sense Galbraith is indeed following in the footsteps of Smith and is inspired by Smith.

9. Indeed, one way to understand Wittgenstein's life work is that he ran into the ubiquitous economist Piero Sraffa and hence, into classical general equilibrium theory. This confrontation induced the dramatic change to the so-called latter Wittgenstein. Economists know from general equilibrium theory that the value of any one good largely depends upon the value of all other goods; and that, generally speaking, there is no such thing as a perfect, invariable, absolute value or measure of value to a commodity. Via Sraffa, Wittgenstein applied these crucial insights of economists to the field of epistemology. The latter Wittgenstein argued that, generally speaking, a single word has no necessary meaning or value in itself. Moreover, the meaning or value of a word depends upon its use and, ultimately, upon the meaning of all other words in the "linguistic field". On the importance of Sraffa to Wittgenstein, see Malcolm (1984: 14–15; 58) and Wittgenstein (1968: X).

10. That contemporary economists should not read Adam Smith may also be Winch's ultimate conclusion (1996: 421).

11. General economists who are not specialists in the history of economic thought, will probably not find the works recommended by Tribe particularly helpful. For example, Winch (1978) is on Smith's politics, not his economics; Winch (1996) writes more on the early reception to Smith's work, rather than on Smith's work itself. Hundert (1994) is mostly about Mandeville – an interesting and important character; but not Smith.

REFERENCES

Brown, V. (1994). *Adam Smith's Discourse. Canonicity, Commerce and Conscience*. London: Routledge.

Coats, W. J. (1994). *A Theory of Republican Characters and Related Essays*. Selinsgrove, N. J.: Susquehanna U. Press.

Colander, D., & Klamer, A. (1987). The Making of an Economist. *Journal of Economic Perspectives, 1*, 95–111.

Evensky, J. (1987). The Two Voices of Adam Smith: Moral Philosopher and Social Critic. *History of Political Economy, 19*, 447–468.

Evensky, J. (1989). The Evolution of Adam Smith's Views of Political Economy. *History of Political Economy, 21*, 123–145.

Evensky, J. (1992). Ethics and the Classical Liberal Tradition in Economics. *History of Political Economy, 24*, 61–77.

Evensky, J. (1993a). Adam Smith on the Human Foundation of a Successful Liberal Society. *History of Political Economy, 25*, 395–412.

Evensky, J. (1993b). Ethics and the Invisible Hand. *Journal of Economic Perspectives, 7*, 197–205.

Evensky, J. (1994a). The Role of Law in Adam Smith's Moral Philosophy: Natural Jurisprudence and Utility. In: R. Malloy & J. Evensky (Eds), *Adam Smith and the Philosophy of Law and Economics* (pp. 199–220). Dordrecht, The Netherlands: Kluwer Academic Press.

Evensky, J. (1994b). Setting the Scene: Adam Smith's Moral Philosophy. In: R. Malloy & J. Evensky (Eds), *Adam Smith and the Philosophy of Law and Economics* (pp. 7–30). Dordrecht, The Netherlands: Kluwer Academic Press.

Evensky, J. (1998). Adam Smith's Moral Philosophy: The Role of Religion and Its Relationship to Philosophy and Ethics in the Evolution of Society. *History of Political Economy, 30,* 17–42.

Evensky, J., & Malloy, R. (Eds) (1994). *Adam Smith and the Philosophy of Law and Economics.* Dordrecht, The Netherlands: Kluwer Academic Press.

Fitzgibbons, A. (1995). *Adam Smith's System of Liberty. Wealth and Virtue: The Moral and Political Foundations of the Wealth of Nations.* Oxford: Oxford U. Press.

Fleischacker, S. (1999). *Judgment and Freedom in Kant and Adam Smith.* Princeton: Princeton U. Press.

Galbraith, J. K. (1971). The Language of Economics. In: J. K. Galbraith *Economics Peace and Laughter* (pp. 32–44). Boston: Houghton Mifflin.

Galbraith, J. K. (1987). *Economics in Perspective. A Critical History.* Boston: Houghton Mifflin Company.

Galbraith, J. K. (1992). *The Culture of Contentment.* Boston: Houghton Mifflin Company.

Griswold, C. L. (1995). Review of V. Brown, *Adam Smith's Discourse: Canonicity, Commerce, and Conscience*; P. Minowitz, *Profits, Priests, and Princes: Adam Smith's Emancipation of Economics from Politics and Religion*; M. J. Shapiro, *Reading 'Adam Smith': Desire, History, and Value*; and P. Werhane, *Adam Smith and His Legacy for Modern Capitalism.* In: *Times Literary Supplement,* July 14, 1995, p. 30.

Griswold, C. L. (1999). *Adam Smith and The Virtues of Enlightenment.* Cambridge: Cambridge U. Press.

Holmes, S. T. (1979). Aristippus in and out of Athens, *American Political Science Review, 73,* 113–128.

Hont, I., & Ignatieff, M. (Eds) (1983a). *Wealth and Virtue.* Cambridge: Cambridge U. Press.

Hont, I., & Ignatieff, M. (Eds) (1983b). Needs and Justice in the *Wealth of Nations*: An Introductory Essay. In: I. Hont & M. Ignatieff (Eds) *Wealth and Virtue* (pp. 1–44). Cambridge: Cambridge Press.

Hundert, E. (1994). *The Enlightenment's "Fable". Bernard Mandeville and the Discovery of Society.* Cambridge: Cambridge U. Press.

Jay, M. (1973). *The Dialectical Imagination. A History of the Frankfurt School and the Institute of Social Research, 1923–1950.* Boston: Little, Brown and Company.

Keynes, J. M. (1964). *The General Theory of Employment, Interest, and Money.* New York: Harcourt Brace Jovanovich.

Lerner, M. (1937). Introduction to *Wealth of Nations,* Cannan Edition. New York: The Modern Library, V-X.

Lux, K. (1990). *Adam Smith's Mistake. How a Moral Philosopher Invented Economics and Ended Morality.* Boston: Shambhala.

Malcolm, N. (1984). *Ludwig Wittgenstein: A Memoir.* Second Edition. Oxford: Oxford U. Press.

McNally, D. (1988). *Political Economy and the Rise of Capitalism. A Reinterpretation.* Berkeley: University of California Press.

Meikle, S. (1995). *Aristotle's Economic Thought.* Oxford: Oxford U. Press.

Minowitz, P. (1993). *Profits, Priests and Princes. Adam Smith's Emancipation of Economics from Politics and Religion.* Stanford U. Press.

Muller, J. Z. (1993). *Adam Smith in His Time and Ours. Designing the Decent Society.* NY: Free Press.

Nichols, J. H. (1979). On the Proper Use of Ancient Political Philosophy: A Comment on Stephen Taylor Holmes's 'Aristippus in and out of Athens'. *American Political Science Review, 73,* 129–133.

Noonan, P. (1990). *What I Saw at the Revolution: A Political Life in the Reagan Era.* New York: Random House.

Ortmann, A. (1999). The Nature and Causes of Corporate Negligence, Sham Lectures, and Ecclesiastical Indolence: Adam Smith on Joint-Stock Companies, Teachers, and Preachers. *History of Political Economy, 31,* 297–315.

Ortmann, A., & Meardon, S. J. (1995). A Game-Theoretic Re-evaluation of Adam Smith's *Theory of Moral Sentiments* and *Wealth of Nations.* In: I. Rima (Ed.) *The Classical Tradition in Economic Thought.* Brookfield, Vt: Elgar.

Ortmann, A., & Meardon, S. J. (1996). Self-Command in Adam Smith's *Theory of Moral Sentiments*: A Game Theoretic Re-interpretation. *Rationality and Society, 8*(1), 57–80.

Pack, S. J. (1991). *Capitalism as a Moral System. Adam Smith's Critique of the Free Market Economy.* Aldershot: Edward Elgar.

Pack, S. J. (1997). Review of Rothbard, *Economic Thought Before Adam Smith: An Austrian Perspective on the History of Economic Thought* and Rothbard, *Classical Economics: An Austrian Perspective on the History of Economic Thought, History of Political Economy, 29,* 367–370.

Pack, S. J. (1998). Murray Rothbard's Adam Smith, *Quarterly Journal of Austrian Economics, 1,* 73–79.

Rashid, S. (1989). Review of Richard F. Teichgraeber, *'Free Trade' and Moral Philosophy. History of Political Economy, 21,* 554–556.

Redman, D. A. (1997). *The Rise of Political Economy as a Science.* Cambridge, MA: MIT Press.

Robinson, J. (1964). *Economic Philosophy.* Harmondsworth: Penguin Books.

Rothbard, M. N. *Economic Thought Before Adam Smith. An Austrian Perspective on the History of Economic Thought. Vol I.* Brookfield, Vt.: Edward Elgar.

Shapiro, M. J. (1993). *Reading 'Adam Smith': Desire, History and Value.* Newbury Park: Sage.

Smith, A. (1976a). *An Inquiry into the Nature and Causes of the Wealth of Nations,* R. H. Campbell & A. S. Skinner (Eds). Oxford: Oxford U. Press.

Smith, A. (1976b). *The Theory of Moral Sentiments,* A. L. Macfie & D. D. Raphael (Eds). Oxford: Oxford U. Press.

Smith, A. (1977). *Correspondence of Adam Smith,* E. C. Mossner & I. S. Ross (Eds). Oxford: Oxford U. Press.

Smith, A. (1978). *Lectures on Jurisprudence,* R. L. Meek, D. D. Raphael & P. G. Stein (Eds). Oxford: Oxford U. Press.

Smith, A. (1980). *Essays on Philosophical Subjects,* W. P. D. Wightman & J. C. Bryce (Eds), Oxford: Oxford U. Press.

Smith, A. (1983). *Lectures on Rhetoric and Belles Lettres,* J. C. Bryce (Ed.). Oxford: Oxford U. Press.

Stigler, G. J. (1975). Smith's Travels on the Ship of State. In: A. S. Skinner & T. Wilson (Eds), *Essays on Adam Smith* (pp. 237–246). London: Oxford U. Press.

Teichgraeber, R. F. (1986). *'Free Trade' and Moral Philosophy. Rethinking the Sources of Adam Smith's 'Wealth of Nations'.* Durham: Duke U. Press.

Thompson, E. P. (1993). *Customs in Common. Studies in Traditional Popular Culture*. New York: The New Press.

Tribe, K. (1999). Adam Smith: Critical Theorist?. *Journal of Economic Literature, 37*, 609–632.

Walker, D. W. (1999). The Relevance for Present Economic Theory of Economic Theory Written in the Past. *Journal of the History of Economic Thought, 21*, 7–26.

Werhane, P. H. (1991). *Adam Smith and His Legacy for Modern Capitalism*. New York: Oxford U. Press.

West, E. G. (1990). *Adam Smith and Modern Economics. From Market Behaviour to Public Choice*. Brookfield, Vermont: Elgar.

West, E. G. (1996). Adam Smith and the Cultural Effects of Specialization: Splenetics versus Economics. *History of Political Economy, 28*, 83–105.

Winch, D. N. (1978). *Adam Smith's Politics*, Cambridge: Cambridge U. Press.

Winch, D. N. (1996). *Riches and Poverty. An Intellectual History of Political Economy in Britain, 1750–1834*. Cambridge: Cambridge U. Press.

Wittgenstein, L. (1968). *Philosophical Investigations*. Oxford: Basil Blackwell.

Young, J. T. (1986). The Impartial Spectator and Natural Jurisprudence: An Interpretation of Adam Smith's Theory of the Natural Price. *History of Political Economy, 18*, 365–382.

Young, J. T. (1990). David Hume and Adam Smith on Value Premises in Economics. *History of Political Economy, 22*, 643–657.

Young, J. T. (1992). Natural Morality and the Ideal Impartial Spectator in Adam Smith. *International Journal of Social Economics, 19*.

Young, J. T. (1995). Natural Jurisprudence and the Theory of Value in Adam Smith. *History of Political Economy, 27*, 755–773.

Young, J. T. (1997). *Economics as a Moral Science. The Political Economy of Adam Smith*. Lyme, New Hampshire: Edward Elgar.

LOVE AND DEATH: THE WEALTH OF IRVING FISHER

Perry Mehrling

ABSTRACT

This essay puts forward a new interpretation of Irving Fisher that integrates his scientific work with his moral crusades, and places both in the context of his times. The key to the new interpretation is Fisher's book on The Nature of Capital and Income *(1906) where he lays out his vision of the economic process and presents his theory of income, neither one of which ever gained acceptance. The new interpretation challenges the standard view of Fisher's scientific work as an anticipation of the post war neoclassical synthesis.*

The light of life only disappears, and its dreary night then commences, when we have none for whom to live. Then the whole creation is a void. Really to live is to live with, and through others, more than in ourselves. To do so we must do so truly. "Love, and love only, is the loan for love". . . .

In so far as to procure good for others, gives a real pleasure to the individual, he is released from that narrow and imperfect sphere of action, to which his mere personal interests would confine him, and the future goods which the sacrifice of present ease or enjoyment may produce, lose the greater part of their uncertainty and worthlessness. Though life may pass from him, he reckons not that his toils, his cares, his privations, will be lost, if they serve as the means of enjoyment to some whom he may leave behind (Rae 1834, 121–122).

There is no question of Irving Fisher's greatness as an economist. Quite the contrary, the evidence of critical commentary indicates that debate persists only

Research in the History of Economic Thought and Methodology, Volume 19A, pages 47–61.

as to what aspect of Fisher's wide-ranging work so qualifies him.[1] Monetarists see *The Purchasing Power of Money* (1911) as the work of a proto-Friedman, while Keynesians see *The Rate of Interest* (1907) and *Booms and Depressions* (1932) as the work of a proto-Keynes. Mathematical economists see Fisher's dissertation *Mathematical Investigations in the Theory of Value and Prices* (1892) as a foundation stone of the theory of general equilibrium subsequently developed by Arrow and Debreu. Empirical economists see *The Making of Index Numbers* (1922) as a foundation stone of the practice of modern government statistical bureaus. In this way, modern divisions within the economics profession have their counterpart in splintered historical interpretation and difficulty with seeing the work entire.

For the most part, commentary limits attention to Fisher's "scientific" work, thus drawing a discrete curtain over that side of Fisher that wanted to tell people *How to Live* (1915), the side of Fisher that promoted hygiene as well as vegetarianism, prohibition, and eugenics. The critical literature seems to agree that we can view these efforts as essentially moral crusades, and trace them to Fisher's childhood as the son of a Congregational minister, a germ source perhaps energized by Fisher's own battle against the tuberculosis that had earlier taken his father. In other words, Fisher's moral crusades can be separated from his scientific work, which means we need have no professional interest in them, scientists as we are. The only crusades that merit our attention are the (scientifically based?) economic crusade for stable money – encompassing successive enthusiasm for the compensated dollar (1920), stamp scrip (1932), and 100% money (1935) – and the somewhat more attenuated campaign for taxation of consumption rather than income (1942).

Similarly, the side of Fisher that tinkered – he invented a tent for tuberculars, a new sun dial, a 13-month calendar, a map projection, and a collapsible chair, as well as the "index visible" that made his fortune (temporarily) – tends to be put aside as reflecting at best an endearing personality quirk, and at worst the psychic need of a man who married wealth to prove himself by generating some of his own. Allen (1993), echoing the sentiments of Schumpeter (1948), speaks for the consensus in bemoaning the loss to economics that resulted from all these peripheral activities. Schumpeter summed up Fisher's scientific work as the "pillars and arches of a temple that was never built" (p. 231) on account of the busywork of the crusader, and opined that Fisher should instead have written a treatise and formed a school of disciples who would take responsibility for interpreting the master and developing his work further. Allen writes: "Had [Fisher] stuck to [his] career as a professional economist and professor, he would certainly have made an even greater contribution and his star would be shining even more brightly in the firmament today" (Allen, 1993, 300).

Although Fisher himself always insisted on the essential unity of all his activities, it has been an operating assumption of most who came after him that Fisher's work doesn't make sense as a whole, so that we are free to pick and choose as we like. An exception is the statistician Max Sasuly, Fisher's associate for twenty years and arguably the closest thing we have to a disciple, who detected "a certain unity and order in [Fisher's] variegated activities . . . the living synthesis in the seeming agglomeration of [his] interests" (1947, p. 257). This essay builds on Sasuly to advance a new more unified interpretation.

Irving Fisher wanted to use science for the betterment of the human condition. To the extent that human problems are economic, Fisher's goal required him to reconstruct economics on a scientific basis, taking the natural sciences as the model of what it means to be a science. This starting point was by no means unique to Fisher and indeed quite well within the spirit of the age in which he lived, but there was less agreement on the nature of science and on what its lessons for economics might be. Fisher's views on these matters turned out to resonate better with the world of post-WWII America than with his own times and in this sense it can be said, as it has been said, that he was a generation ahead of everyone else. But to really understand him, we need to approach him on his own terms in the context of his own time, rather than on our own terms and in retrospect.

THE STANDARD VIEW: THE THEORY OF VALUE AND PRICES

For a modern economist, it is easiest to gain access to Fisher by starting with his dissertation (1892) which treats the problem of relative price determination in general equilibrium with maximizing consumers. For the most part the mathematical structure is familiar (leaving aside the treatment of production) so we feel that we understand him immediately. When we find much the same mathematical structure in *The Rate of Interest* (1907), albeit relegated to a mathematical appendix, we are naturally inclined to see the book as the intertemporal extension of the timeless general equilibrium model of the dissertation. Such an interpretation is made all the more compelling by Fisher's graphical presentation which is familiar from any intermediate microeconomics textbook. We find utility maximization subject to a budget constraint (p. 387), income maximization given production possibilities (p. 402), and general equilibrium of production and consumption choice (p. 409). Fisher must be like us.

From this starting point, much of the rest of Fisher's opus falls into a certain order. Most important, it is apparent already in his dissertation that Fisher is a natural quantity theorist because he sees money first and foremost as a unit of measurement. Writes Fisher: "Money is here used solely as a measure of value. It is not one of the commodities in the market. The high or low price of commodities in terms of this money is dependent entirely on the amount of it at which we agree to rate the yearly consumption of the market" (1892, 41). Ten pages later, discussing the comparative statics of the model, he drives the point home: "Increase all incomes in the same ratio. Then will all prices increase and the valuation of money decrease exactly in this ratio. There will be no change in the distribution of commodities. There is merely a depreciated standard of money" (1892, 51). Here certainly is the origin of Fisher's later *Purchasing Power of Money* (1911).[2]

Not only is Fisher a natural quantity theorist, but in the dissertation he is already thinking along the lines that will lead to his idea (and subsequent crusade) for the compensated dollar (1911, Ch. 13). Already he sees the gold standard as a primitive attempt to tie down the standard of value (1892, 58), primitive for the consequence that all prices must change whenever the relative price of gold changes. And already he sees a better way: "it is perfectly possible to have a measure of value which is not a commodity at all. Thus we might agree to call the consumption of the United States for a year $10,000,000,000, and this agreement would immediately fix a measure of value, though the new dollar need have no equality to the gold or silver dollar. It would be easy to translate between such an arbitrary standard and any commodity standard. Thus if statistics showed that the consumption measured in gold dollars was $12,000,000,000, the agreed standard is at 120 compared with gold and by measure of this factor we can reduce the prices of all commodities" (p. 41). Here certainly is the origin of Fisher's idea to stabilize prices by creating a new dollar whose value against gold would be changed depending on an index of the general price level.

So much for the static equilibrium side of Fisher, the dynamic disequilibrium side is also presaged in the dissertation. "The ideal statical condition assumed in our analysis is never satisfied in fact" (p. 103) but that is just because "the dynamical side of economics has never yet received systematic treatment" (p. 104). It is easy enough to see here the origin of Fisher's later discussion of "transition periods" (1911, Ch. 4), as well as the Fisher who saw business cycles as nothing more than the "Dance of the Dollar" (1923) caused by a lagged response of nominal interest rates to monetary inflation and deflation.

The Fisher who in 1892 wrote that "Panics show a lack of equilibrium" (p. 103) is recognizably the Fisher who proposed "The Debt-Deflation Theory of Great Depressions" (1933). From the beginning to the end, Fisher always had in mind the concept of equilibrium as reference point and central tendency of the economy, but only ever as an approximation to actual conditions. He thought monetary reform would help keep us closer to the theoretical equilibrium, since monetary fluctuation is apparently the cause of much of the disequilibrium phenomena we observe, and this explains Fisher's lifelong devotion to the cause of stable money.

It's a nice story. It puts most of the pieces in place and, *mirabile dictu*, the picture that emerges is the postwar neoclassical synthesis! The broad church of *Fishergeschichte* turns out to be just the broad church of modern economics. True, there is no room in this story for Fisher the Crusader, but we can all celebrate Fisher the Scientist, the young mathematician who got interested in economics, saw a system amenable to analysis as if it were a physical system, and wrote in one year the dissertation that set the course of economics for the next century. Well, not all of us can celebrate, since some regret the consequent diminished role for history in the new theoretical science. Thorstein Veblen saw it happening and spoke up at the moment (Veblen 1908, 1909); Veblen's modern representatives follow suit (Mirowski 1989). But even these critics essentially agree with the consensus interpretation of what Fisher did; they just don't like it. They are the exception that proves the standard view of Fisher.

An interpretation that fits so many of the facts must be right, at least in part. But there is at least one piece of the puzzle that doesn't fit so well and that points to another line of interpretation. That piece is the first book Fisher wrote after recovering from tuberculosis, *The Nature of Capital and Income* (1906). Conventional interpretation treats the book as a kind of preliminary for *The Rate of Interest* (1907) and, since *The Rate of Interest* evidently extends the dissertation to intertemporal general equilibrium, by association *Nature of Capital and Income* is presumed to tie in with the dissertation in some way. The problem is that the tie is never explored. Indeed, the standard interpretation of *The Rate of Interest* leaves us wondering why it needs any preliminary. After all, in modern economics we pass immediately from the exchange economy to the intertemporal case, merely by relabeling the axes. If Fisher is just doing this modern trick for the very first time, why did he feel the need for any pause in between, much less a book-length pause? And, on the subject of the pause, remember that this is the book he wrote twice, since the only copy of the manuscript was lost to the thief who stole Fisher's briefcase when he put it down for a moment to make a phone call in Grand Central station (Allen, 1993, 93; Fisher, 1956, 102, 126).

Big pause, big book. *The Nature of Capital and Income* is where Fisher developed his distinct concept of income, a concept moreover that he always insisted was absolutely central to his theory of interest. Modern economics has adopted Fisher's theory of interest but remained skeptical about his concept of income, thus picking and choosing among what Fisher always insisted were inseparable components of a unity. Maybe the easiest entry point into Fisher's work is just a bit too easy. Maybe Fisher is not quite as much like us as first appears.

A SHIFT OF FOCUS: CAPITAL, INCOME, AND TIME

Take then *The Nature of Capital and Income* as an alternative starting point for interpretation. The book appears to be a treatise on accounting, and on that score we might be inclined to discount its importance, but that is a mistake. In dealing with accounting, Fisher is dealing with the world of practical business, a world at that time (as now?) quite distant from the world of economic theory. In his dissertation he had deployed techniques and habits of thought learned from mathematical physicist Willard Gibbs in order to construct a theory of economic equilibrium, but he really didn't know very much economics yet, only what he had learned from courses with William G. Sumner ("who first inspired me with a love for economic science" according to the dedication of *Capital and Income*). The 1906 book takes an approach quite different from the dissertation, looking instead to deploy techniques and habits of thought from the world of practical business. If we read the book on its own, what comes through is nothing less than an overarching vision of the economic process,[3] and that vision is *not* what we find in standard microeconomics textbooks. It is closer to the vision we find in the musings of modern financiers such as George Soros (1987) or Henry Kaufman (2000), or of academics who made it a practice to talk with and listen to such financiers, academics such as Hyman Minsky (1986) or Fischer Black (1987).

The classical economists habitually thought of the present as determined by the past. In Adam Smith, capital is an accumulation from the careful saving of past generations, and much of modern economics still retains this old idea of the essential scarcity of capital, and of the consequent virtue attached to parsimony. A financier tends to view these matters in a different light. From a financial point of view, the present is determined by the future, or rather by our ideas about the future. Capital is less a thing than an idea about future income flows discounted back to the present. For a financier, the emphasis is thus on how new ideas, not new savings, augment and shift current capital values.

In *The Nature of Capital and Income*, Irving Fisher straddles the worlds of classical economics and practical finance by distinguishing physical capital goods (for which the past-determines-present view is correct) from the value of those goods (for which the future-determines-present view is correct).[4] Fisher has a foot in both worlds, but we can tell where his heart is since he defines income as the realized flow of services that, when capitalized, gives rise to current capital value. In his formulation, the current valuation of capital takes into account all future income expected to be generated by capital, so anything that changes expectations changes also current capital values. It follows that capital value can be augmented without any capital accumulation whatsoever, but we don't want to count such capital gains as current income since it is actually a change in (expected) future income and we don't want to double count. Furthermore, it follows that even capital accumulation (saving) we don't want to count as part of current income since it also will show up in our accounts as a change in capital value on account of the future income to which it is expected to give rise. Here too, we want to be careful not to count the same income twice.

How is this treatise on accounting a vision of the economic process? The accounts present a unified picture of the economy as a whole as nothing more than a stock of wealth moving through time, throwing off a flow of services as it goes. (Note well, *not* a static exchange economy.) In Fisher's formulation all wealth is capital, not just machines and buildings, but also land and even human beings. Indeed for Fisher human beings are the most important form of capital because the most versatile. Thus, at the highest level of abstraction, there is no distinction between the traditional categories of labor, capital, and land. All produce a stream of income (services) so all are capital, and their future income discounted back to the present is their capital value. Similarly, at the highest level of abstraction, there is no distinction between the traditional categories of wages, profit, and rent. All are incomes thrown off by capital, hence all are forms of the more general category of interest. From this point of view, Fisher's next book *The Rate of Interest* (1907) appears *not* as a theory of intertemporal price equilibrium, but rather as a theory of the rate of income flow at a moment in time. The rate of interest is not a price but a yield.

The book also appears as a theory of social welfare. The value of wealth, so broadly conceived as Fisher intends, is an operational measure of welfare since anything that increases the value of wealth must necessarily do so by increasing the flow of final services enjoyed by human beings.[5] Thus Fisher's concept of wealth gave him a concrete quantitative framework to guide his normative impulses, as will be seen presently. It also gave him, and this needs to be brought out first, a quantitative way of understanding the course of economic progress up to his own time. In this sense, *The Nature of Capital and Interest* needs to be understood as the prolegomenon to a theory of economic development that Fisher never actually wrote (hence Schumpeter's disappointment).

More insight into this latter strand of Fisher's thought can be gleaned by an appreciation of its apparent origin in John Rae's *New Principles on the Subject of Political Economy* (1834). In February 1897 Fisher published a brief note on Rae in the *Yale Review*. In November 1905 he reviewed a reissue of Rae's book that had been put together by C. W. Mixter and republished as *The Sociological Theory of Capital* (1905). The intervening years were the years of Fisher's tuberculosis, a fallow period in his intellectual life.[6] Fisher subsequently went on to dedicate *The Rate of Interest* (1907) to Rae ("To the memory of John Rae who laid the foundations upon which I have endeavored to build"). When he published a revised version of the book as *The Theory of Interest* (1930), he added Bohm-Bawerk to the dedication but he also wrote in the preface, "Every essential part of it was at least foreshadowed by John Rae in 1834" (p. ix). There is plenty of evidence that Rae influenced Fisher, but what exactly was the nature of that influence?

Rae's book is essentially a theory of economic development presented as a critique and alternative to the theory of Adam Smith. As against Smith's emphasis on accumulation through parsimony, Rae emphasizes that individuals may accumulate wealth without the nation becoming any wealthier if their wealth is merely acquired from others rather than being newly created. Thus the laissez faire recommendation to allow individuals free rein to accumulate individual wealth may not yield the desired social results. He concludes that the emphasis should rather be on invention which, though it may or may not increase individual wealth, certainly creates new wealth at the level of society as a whole. Such an emphasis, Rae goes on to argue, leaves considerable room for state intervention, by contrast to the laissez faire program that emerges from Smith's theory. In this way, Rae offered Fisher a conception of economic development that fit with his own emerging conceptions of capital and income. And he also offered Fisher what his revered teacher Sumner never did, a vision of the economic process that leaves room for its improvement.

According to Rae, the "legislator" can increase "national stock" or wealth by (1) promoting the general intelligence and morality of the society, (2) promoting invention, and (3) preventing dissipation of national wealth in luxury (Rae 1834, 362). The first is directed toward increasing the desire to accumulate, the second toward increasing the opportunity to accumulate, and the third toward making sure that accumulated wealth is not lost. Viewed in this light, Fisher's theory of interest can be understood not as a celebration of the optimality of market outcomes but rather as a framework for thinking about how to increase the flow of future services and hence current wealth. Here is Fisher in 1930: "From what has been said it is clear that, in order to estimate the possible variation in the rate of interest, we may, broadly speaking, take account of the following three groups of causes: (1) the thrift, foresight, self-control, and love of offspring which exist in the community; (2) the progress of inventions; (3) the changes in the purchasing power of money" (p. 515). This evidently parallels Rae, with the exception of the third point, but now we can understand something more about that point as well. If aggregate wealth is ever to serve as an operational welfare measure, it is important that the rod we use to measure wealth is itself stable in value. Quite apart from causing business fluctuation, monetary instability makes it hard for us to know whether a measured increase in wealth is actually an increase in aggregate welfare or only an increase in average prices.

This interpretation brings Fisher's various moral crusades into line with his scientific work. The standard interpretation suggests that Fisher's crusades were sparked by his tuberculosis, but a case can be made that it was not the tuberculosis but rather what Fisher was reading while he was recovering from the tuberculosis, namely Rae, that sparked the change. From this point of view, it appears that not just his monetary crusades but all of them stem from a scientific basis as Fisher understood it. Barber (1997) has pointed out that Fisher's crusades were all about maximizing efficiency and eliminating waste, and this seems right so far as it goes, but it doesn't go far enough. Every one of Fisher's presumed "moral" crusades can be seen as the action of Rae's "legislator" looking to increase the national stock.

Thus, to educate people about the "Rules for Healthful Living Based on Modern Science" (the subtitle of *How to Live*) – there were fifteen rules, ranging from fresh air and good food to regular bowel habits and cultivation of mental serenity – was not only or even mainly about making them better people. For someone who sees human beings as part of the stock of national wealth, as Fisher did, even a few years added to the effective working life of each person amounts to a tremendous increase of current national wealth. Similarly, Fisher's raving about eugenics, the bit of Fisher guaranteed to make a modern audience most uncomfortable, begins to come into sharper focus. When Fisher insists that "this germ plasm, which we receive and transmit, really belongs, not to us, but to the race. . . . [and] we are under the most solemn obligation to keep it up to the highest level within our power" (1915, 165), he is talking as an economist about the national stock of wealth as he understands it. Finally, Fisher the tinkerer fits too. Rae's emphasis on the importance of invention shows up not only in Fisher's scientific work (1907 Ch. 10; 1930 Ch. 16), but also in the tinkering that made him a fortune during the 1920s, and in the financial investment practices that lost him that fortune in the 1930s. By helping, as he thought, to foster invention he was only doing his own bit to maximize social welfare by increasing aggregate wealth.

Given the impact that Rae seems to have had on Fisher, it is interesting to note two areas in which Rae's teaching did not penetrate. First, science. Rae contrasts Adam Smith's axiomatic philosophical approach unfavorably to the scientific inductive empiricism of Francis Bacon, and he claims the latter method as his own (Rae, 1834, 328–333). In Rae's view, the true scientific method involves piling fact upon fact in order to find eventually the true laws of nature. We can see Fisher bowing in that direction in *Rate of Interest* when he includes three final chapters offering "inductive verification" of his own theory (Ch. 14–15) and "inductive refutation" of the monetary theory of interest (Ch. 16). But, truth told, none of the chapters really amounts to empiricism in the Baconian spirit. It all seems more like marshaling evidence to illustrate a prior theoretical conclusion. Fisher's subsequent *Report on National Vitality* (1909) comes closer to the empiricist method, but even here there is the distinct flavor of marshaling evidence for a conclusion decided on other grounds. Rae may have opened Fisher's eyes to the value of the research being done by those who were proceeding more in the Baconian spirit (for example, Wesley Clair Mitchell) but in the end the process of piling one fact on another was too slow for Fisher. He hoped to do better by adopting a more theoretically directed approach, one informed both by economic and statistical theory.[7]

Second, money. Rae was an enthusiast for the Scottish system of free banking (Rae, 1834, 176–193, 397–412). He wrote: "The real advantage of the art [of banking] arises from its application of the floating loans of the society to the purposes of exchange" (p. 412). None of this sunk in. On the subject of money, Fisher stuck with Sumner's quantity theoretic explanation of prices given an inconvertible currency (Sumner, 1874, 221) and also with Sumner's consequent admiration for the British innovation of keeping the currency issue fixed in the Issue Department "entirely separated from all the vicissitudes of the banking business" (Sumner, 1896, 465). Fisher stuck with Sumner's theoretical understanding in part, one gathers, because it appeared to be consistent with his own 1892 scientific researches. He departed from Sumner not on theory but on application, favoring managed money rather than the laissez faire gold standard.

CONCLUSION

In his 1910 textbook, Fisher wrote: "I am one of those who believe that when the usage of academic economics conflicts with the ordinary usage of business, the latter is generally the better guide" (1910, xiv). We have seen how he followed this precept in *The Nature of Capital and Income* with good results in terms of connecting up the practice of accounting with the concepts of economics. We have also seen, however, that he did not follow this guide when it came to money, where his mind was already made up by the time of his dissertation, rendering him impervious even to the suggestive comments of his hero Rae. Fisher knew about monetary theories of the rate of interest, and he knew that they emerged from the ordinary usage of business. He rejected them anyway. Fisher thereby founded modern finance on the principle that liquidity effects can be ignored for the purposes of constructing an equilibrium theory of finance, and he founded modern macroeconomics on the principle that monetary fluctuations have only temporary disequilibrium effects.

Whatever one may think of what he did, it is hard to see how Fisher could have done otherwise, given who he was and what he was trying to do. Listening to the bankers (as for example did Laurence Laughlin, the anti-quantity-theory Chicago economist) would have meant giving up the concept of income that Fisher needed for thinking about social welfare. In a thoroughgoing money flow view of the world, much of what Fisher's income concept brings into focus seems unimportant (for example, the implicit flow of services from consumer durables), and much of what Fisher's concept pushes into the background seems highly important (for example, financial flows). Fisher wanted a concept of income and capital that he could sum, because he wanted to measure aggregate or social income. For a money flow view, such summation is out of the question since it immediately nets out the phenomena that need to be front and center in the analysis. No, Fisher had to shut his ears to the practical bankers if he was ever going to make progress on his own agenda.

One consequence was that Fisher played no constructive role in the establishment of the Federal Reserve System in 1913. Another consequence was that Fisher played no constructive role in the attempts then underway by a small group of academics to develop the money flow view into a proper theory of money. I'm thinking here of Veblen (1909) but also of Young (1911) and Mitchell (1910, 1916) whose writings need to be read as responses to Fisher and attempts to articulate a positive alternative.[8] They could have used the help of a mind like Fisher's, but it was not to be.

Fisher had his own way of thinking about these matters and that, more or less, was that. He had no interest in revisiting fundamental issues that he felt he had resolved. And he couldn't help thinking that somehow his critics just weren't ready for a truly scientific approach to these questions, or didn't adequately appreciate the enormous potential benefits to mankind toward which he was working. Fisher's response to Veblen's dismissive review is a case in point, evincing equal parts puzzlement and determination to proceed: "There are doubtless many points of difference between us, but they are not in general those which Professor Veblen has mentioned. Unless unwittingly I do him injustice, his preconceptions have led him to misconceive my method and conclusions and to confuse them with methods which we both oppose" (1909, p. 516). Misconceptions there most definitely were, but on both sides and, one is bound to conclude, inevitably if somewhat tragically so.

By the time Fisher published *The Rate of Interest*, he had more or less finished his life work of theoretical construction. It was not the kind of temple Schumpeter would have liked to see, but it was a temple nonetheless, not just pillars and arches. From then on Fisher turned his attention to application, expansion, and persuasion. Schumpeter regretted the loss to economics, but Fisher himself never did. For him these activities were not peripheral. They were the whole point of his science.

NOTES

1. The principle biographical sources are Irving Norton Fisher (1956) and Robert Loring Allen (1993). The most influential interpretations appear to be Schumpeter's obituary essay (1948), the papers in the collection edited by Fellner (1967), and Tobin's (1985) retrospective. A sample of the most recent interpretation is collected is Loef and Monissen (1999). The interpretation offered in the present essay builds most directly on the suggestive and neglected essay of Sasuly (1947).

2. Fisher's famous distinction between nominal and real interest appears to have its origin not in the dissertation but in Fisher's later exploration of the economics of bimetallism. *Appreciation and Interest* (1896) is mainly concerned with the relationship between the rate of interest under a silver standard as opposed to a gold standard. Fisher understands both rates as the sum of an underlying "real" rate plus appreciation or depreciation of the money standard in question. Only later did he apply the same reasoning to a flat money standard in order to arrive at the familiar formula.

3. Schumpeter (1948, 223) suggests as much, but does not characterize the vision that emerges.

4. Keynes might be understood as having been engaged in the same straddle. This helps to explain the attraction that American Keynesians (e.g. Tobin, 1985) felt for Irving Fisher. And it helps to explain the attraction that young American economists, already exposed to Fisher, felt for Keynes. Note however the difference between Fisher

and Keynes on the definition of income. Also, whereas Keynes mainly used uncertainty to get time into the picture, Fisher mainly used invention, and he did so even though he was familiar with the distinction between risk and uncertainty. (Compare Fisher, 1906, Ch. 16 "The Risk Element" with Keynes' *Treatise on Probability* 1921.)

5. It is a crude measure, of course, and Fisher thought that direct utility measurement was still needed in order to take account of cross-section variation in the marginal utility of income. Fisher's 1927 article on "Measuring Marginal Utility" is an attempt to take that next step.

6. Allen (1993) dates the tuberculosis episode from diagnosis in fall 1898 to recovery of full energy in spring 1904 after the fire that destroyed Fisher's house at 460 Prospect Street.

7. Sasuly (1947) is a good starting point for understanding this later side of Fisher. In the recent collection by Loef and Monissen (1999), the essays by John Chipman on Fisher's econometrics and by Janos Barta on his index numbers are worth close attention.

8. Mehrling (1997) chronicles the line of thinking about money that springs from these origins. Note particularly the role of J. B. Canning's *Economics of Accountancy* (1929) as a formative influence on Edward Shaw, an influence later incorporated in his book with John Gurley *Money in a Theory of Finance* (1960). Canning's book also influenced Irving Fisher. In rewriting *The Rate of Interest* (1907) as *The Theory of Interest* (1930), Fisher explicitly followed Cannings' suggestion to put the flow of income at the center of the analysis, and the stock value of capital more in the background. It's not the money flow view, but it moves as much in that direction as was possible for Fisher.

REFERENCES

Allen, R. L. (1993). *Irving Fisher, A Biography.* Cambridge, Mass.: Blackwell.

Barber, W. J. (1997). Career Highlights and Formative Influences. In: W. J. Barber (Ed.) *The Works of Irving Fisher* (pp. 3–21). London: Pickering and Chatto.

Black, F. (1987). *Business Cycles and Equilibrium.* New York: Basil Blackwell.

Canning, J. B. (1929). *The Economics of Accountancy: a critical analysis of accounting theory.* New York: Ronald Press.

Fellner, W. (Ed.) (1967). *Ten Economic Studies in the Tradition of Irving Fisher.* New York: Wiley.

Fisher, I. (1892). *Mathematical Investigations in the Theory of Value and Prices.* PhD Thesis. Transactions of the Connecticut Academy, *9* (July): 1–124.

Fisher, I. (1887). A Neglected Economist. *Yale Review, 5*(4), (February), 457.

Fisher, I. (1896). *Appreciation and Interest.* Publication of the American Economic Association, Third Series, Vol. XI, No. 4. New York: Macmillan.

Fisher, I. (1905). Review of *The Sociological Theory of Capital* by John Rae. *Yale Review, 14*(3), (November), 330–333.

Fisher, I. (1906). *The Nature of Capital and Income.* New York: Macmillan.

Fisher, I. (1907). *The Rate of Interest; Its Nature, Determination and Relation to Economic Phenomena.* New York: Macmillan.

Fisher, I. (1909). *Report on National Vitality, Its Wastes and Conservation*. Washington, D.C.: Government Printing Office.

Fisher, I. (1909). Capital and Interest. *Political Science Quarterly, 24*(3), (September), 504–516.

Fisher, I. (1910). *Elementary Principles of Economics*. New York: Macmillan.

Fisher, I. (1911). *The Purchasing Power of Money; Its Determination and Relation to Credit, Interest and Crises*. New York: Macmillan.

Fisher, I. (1915). *How to Live; Rules for Healthful Living Based on Modern Science* (with Eugene Lyman Fisk). New York: Funk and Wagnalls.

Fisher, I. (1920). *Stabilizing the Dollar*. New York: Macmillan.

Fisher, I. (1922). *The Making of Index Numbers*. Boston: Houghton Mifflin.

Fisher, I. (1923). The business cycle largely a 'dance of the dollar'. *Journal of the American Statistical Association, 18*, 1024–1028.

Fisher, I. (1927). A statistical method for measuring marginal utility and testing the justice of a progressive income tax. In: J. H. Hollander (Ed.), *Economic Essays, Contributed in Honor of John Bates Clark* (pp. 157–193). New York: Macmillan.

Fisher, I. (1930). *The Theory of Interest, as determined by impatience to spend income and opportunity to invest it*. New York: Macmillan.

Fisher, I. (1932). *Stamp Scrip*. New York: Adelphi.

Fisher, I. (1932). *Booms and Depressions: some first principles*. New York: Adelphi.

Fisher, I. (1933). The Debt-Deflation Theory of Great Depressions. *Econometrica, 1*, 337–357.

Fisher, I. (1935). *100% Money*. New York: Adelphi.

Fisher, I. (1942). *Constructive Income Taxation*. New York: Harper.

Fisher, I. N. (1956). *My Father, Irving Fisher*. New York: Comet Press.

Gurley, J. G., & Shaw, E. S. (1960). *Money in a Theory of Finance*. Washington, D. C.: Brookings Institution.

Kaufman, H. (2000). *On Money and Markets*. New York: McGraw Hill.

Keynes, J. M. (1921). *A Treatise on Probability*. London: Macmillan.

Loef, H.-E., & Monissen, H.G. (Eds) (1999). *The Economics of Irving Fisher, Reviewing the Scientific Work of a Great Economist*. Northampton, Mass.: Edward Elgar.

Mehrling, P. (1997). *The Money Interest and the Public Interest, American Monetary Thought, 1920–1970*. Cambridge, Mass.: Harvard University Press.

Minsky, H. (1986). *Stabilizing an Unstable Economy*. New Haven: Yale University Press.

Mitchell, W. C. (1910). The rationality of economic activity. *Journal of Political Economy, 18*(2–3), (February, March), 97–113, 197–216.

Mitchell, W. C. (1916). The role of money in economic theory. *American Economic Review, 6*(1), supplement (March), 140–161.

Mirowski, P. (1989). *More Heat Than Light: Economics as social physics, physics as nature's economics*. Cambridge, England: Cambridge University Press.

Rae, J. (1834). *Statement of Some New Principles on the Subject of Political Economy, exposing the fallacies of the system of free trade*. Boston: Hilliard, G.

Sasuly, M. (1947). Irving Fisher and Social Science. *Econometrica, 15*(4), (October), 255–278.

Schumpeter, J. A. (1948). Irving Fisher's Econometrics. *Econometrica, 16*, 219–231.

Soros, G. (1987). *The alchemy of finance: reading the mind of the market*. New York: Simon and Schuster.

Sumner, W. G. (1874). *A History of American Currency*. New York: H. Holt.

Sumner, W. G. (1896). A History of Banking in the United States. New York: *Journal of Commerce and Commercial Bulletin*.

Tobin, J. (1985). Neoclassical Theory in America: J. B. Clark and Irving Fisher. *American Economic Review, 75*(6), (December), 28–38.

Veblen, T. (1908). Fisher's Capital and Income. *Political Science Quarterly, 23*(1), (March), 112–128.

Veblen, T. (1909). Fisher's Rate of Interest. *Political Science Quarterly, 24*(2), (June), 296–303.

Young, A. A. (1911). Some limitations of the value concept. *Quarterly Journal of Economics, 25*(3), (May), 409–428.

THE ECONOMY OF
JAMES BRANCH CABELL

Warren J. Samuels

ABSTRACT

James Branch Cabell was an American journalist, novelist, and essayist,
whose best-known work is that of fantasy. His 1919 work, Beyond Life, has
a chapter entitled "Which Admires the Economist." In it he explores three
modes of or attitudes toward life, together constituting the economy: the
gallant, the chivalrous, and the poetic, the last of the three providing the
raw materials for creativity. The exchange economy is combined with the
creative versus the prosaic person and with the abstract versus the
practical modes of living. The result amounts, in part, to a theory of
entrepreneurship or leadership.

Economists have identified economics in many ways over the centuries. The
domain of economics has been defined, inter alia, in terms of production,
exchange and consumption; the production and amassing of wealth; the
ordinary business of life, namely, earning a living; the allocation of resources;
the arrangements established for the provisioning of human needs; the process
of dealing with scarcity; the market and/or price system; and the logic of
choice. One type of approach contemplates a particular domain designated as
the economy. Another type projects an economic aspect of all spheres of life.

Research in the History of Economic Thought and Methodology, Volume 19A,
pages 63–73.
© 2001 by Elsevier Science B.V.
ISBN: 0-7623-0703-X

All such delimiting identifications and definitions need not – and perhaps do not – correspond to anything ontologically "real", representing instead conceptual and methodological tools with which to organize and channel serious thought.

Perhaps the most widespread contemporary view, albeit one not without controversy, is owed, in a manner of speaking, to Lionel Robbins, late of the London School of Economics, and to Gary Becker, currently of the University of Chicago and one of the leaders of the Chicago School of Economics. The difference between the London and the Chicago "schools" is one of definition, the London School being a university and the Chicago School, while nominally located in a university (and with members elsewhere), is a group of economists sharing certain common doctrines. At any rate, the Robbins-Becker view is that economists do not study "the economy" but rather the economic aspect of all of life, centering on the necessity of choice due to scarcity of resources which have alternative uses. The economic aspect is conceptualized as that which involves individual rational calculation, constrained maximization, the weighing and comparing of benefits and costs, . . . in short, a utilitarian rather than a customary or romantic or sociological or political point of view.

This view can be held in at least two ways. In one, the presumption is that the story being told is one of how people actually think and behave. It it intended to be a definition of reality. In the other, the story is only the result of the use of a set of limiting assumptions – tools – deemed useful for examining certain aspects of life "as if" people actually thought and behaved that way, without any presumption that they do so. (So, too, could other views, ones focusing on the roles of power, the working rules of law and morals, the totality of institutional complexes, and so on.)

The view that economists study the economic aspect of all of life has facilitated what has been called – with praise and with derision – "economics imperialism", the effort to apply the categories of economic theorizing to, and perhaps to capture, other disciplines, especially political science, law and sociology. This approach, in substantially all its uses, is much more credible as an engine of inquiry and insight than a self-contained exclusivist explanation. It is most useful in exploring interesting facets of behavior but of doubtful reliability as an indicator of its fundamental nature or as explanation. Nonetheless, its use tends to be largely of the latter type.

Interestingly, several non-economists, such as Charles Saunders Peirce and Sigmund Freud, made occasion to use the concept of an economic aspect in their work.[1] Another noneconomist who did so was James Branch Cabell who, in his 1919 literary work *Beyond Life*, entitled Chapter V as the one "Which

Admires the Economist".[2] Cabell's use, actually uses, showed the concept's power as an engine of inquiry and insight.

* * * * *

James Branch Cabell (1879–1958) was an American journalist, novelist and essayist.[3] Cabell had a negligible reputation until his prosecution in 1920 for obscenity (in his 1919 novel *Jurgen*) by the New York Society for the Prevention of Vice. Acquitted of the charge, he gained national celebrity and something of the status of a literary cult figure. His reputation waxed and waned, most recently remembered but not extolled except for a few, like Carl Van Doren, Edward Wagenknecht, Edmund Wilson, Joe Lee Davis and Louis Untermeyer, who gave him a loftier ranking among American literary figures. Much of his best known work is that of a fantasist, ranked by some with the creations of John R. Tolkien and Richard Brautigan, especially his eighteen-volume *The Biography of the Life of Manuel*. A graduate of William and Mary College in 1898, Cabell worked for the New York Herald and Richmond News between 1899 and 1901, then struggled until 1920 as an author of magazine stories and a dozen novels and as a genealogist. Thereafter, he published some three dozen books, the last an autobiography in 1955. In addition, many extensively revised editions were published as well as a major collection of his chief works. Cabell also was an editor of *The American Spectator* during 1932–1935.

The critical work on Cabell points to his skepticism and pessimism, his ridicule of human beings' belief in their self-importance, and his more or less grudging affirmation of a life of ordinary contentment. Looked at somewhat differently, Cabell is interpreted as emphasizing man as a creature of his dreams, courting both nobility and absurdity, but persevering in the power to believe in the constructs of his own imagination, however illusory. In Cabell's view, the imaginary or the dreamed is for man the true reality. It is believed that Cabell was reacting to philosophical and literary naturalism, rejecting the mere recording of the observable as incomplete, lacking the meaning ascribed by human imagination.

Pertinent to the themes constituting the economy of James Branch Cabell are what he identified as the three modes of or attitudes toward life: the gallant, regarding life as play; the chivalrous, regarding life as a sacred trust; and the poetic, regarding life as the raw materials for creation. One can imagine an economy in which these three attitudes, and their attendant lifestyles, are traded off one for the other. But one sensitive critic affirms that, "for all its technical and narrative richness", the "basic premises" of Cabell's art "are simple: that human living is everywhere pretty much the same, but that life is blessedly fertile in disguises and that in most cases even the most unconventional and defiant of human quests lead round about to quite predictable and ordinary ends, which fact, no doubt painful to the romantic and the poet, provides the perceptive with a good deal of quiet amusement and is something of a comfort to the ordinary citizen. On these few premises hangs the whole intricate structure" (Wells, 1981, p. 113). The same author notes that Cabell, a stubborn individualist possessed of his "own highly idiosyncratic vision" in matters of aesthetics and literary theory and practice, had a "reputation as an iconoclast and a sly circumventer of taboos" (Wells also thinks that Cabell's iconoclasm can be exaggerated, that it is "a relatively insignificant feature of Cabell's work" (Wells, 1981, p. 115)). One wonders about the relation of the foregoing premises to both his own attitudes and his life.

* * * * *

Cabell's chapter four, "Which Admires the Economist", uses an interpretation, characterization and evaluation of some of the works of Christopher Marlowe (1564–1593) to identify "the economy", or, more properly, the economic aspect of life. Among other things, Cabell compares Marlowe with William Shakespeare (1564–1616) and François Villon (1431–1463(?)), somewhat of a precursor, in mode of living, of Jean Genet) and discourses on the creative role of drugs and alcohol (about which more anon).

Cabell rejects the thinking of "those short-sighted persons who somehow confound economy with monetary matters (p. 86). Alas, nowhere does Cabell directly elaborate on his own meaning of economy, but it is clear he finds it pertinent to understanding the work of creative people, especially writers like Marlowe. His meaning becomes clear through his interpretation of Marlowe's work.

Marlowe, he writes, is not simply a poet. Marlowe's "real daring, like that of all the elect among creative writers, was displayed as an economist. And it is the economy of such poets", he points out, "that I must pause to explain" (p. 86). Indeed, "It is unfair that ... Marlowe should be accorded no very general consideration as an economist" (p. 89). He thus writes of "the economy of Marlowe" (p. 92), "Marlowe's economy" (p. 101), and his (and Villon's) "talent for economy" (p. 110).

Cabell's substantive analysis is presented in three steps, each more sophisticated than the preceding one(s).

First, Cabell elicits the exchange nature of economy from Marlowe's *The Tragical History of Dr. Faustus*. Here we have the original Faustian bargain, "the legend [as Cabell summarizes it] of the sorcerer who, in exchange for his soul, leased of the devil Mephistophilis a quarter-century tenure of superhuman powers, and at the running out of his bond was carried off alive to hell" (pp. 98–99). The exploit in the legend which, Cabell argues, "most deeply impressed Marlowe was the evocation of Helen of Troy, in defiance of time and death, and any process of human reason, to be the wizard's mistress" (p. 99). But exchange involves the payment of a price: "a man must pay dearly for doing – not what heaven disapproves of, as would speed the orthodox tag – but that which heaven nowadays does not permit" (p. 100). The price is not felt until its payment is due: "There is really no trace of regret for the hellish [a wonderful pun] compact until punishment therefor impends" (pp. 100–101).

Second, Cabell deploys an argument which is very near and dear to him, one which contrasts the creative person, who seeks permanent creations for the ages, to the prosaic person, who lives a life of unmitigated ordinariness.[4] The former person, who lives to create, is extolled. In an earlier chapter, Cabell writes (in words some of which have been widely quoted and which is very deep psychologically),

> For thus to spin romances is to bring about, in every sense, man's recreation, since man alone of animals can, actually, acquire a trait by assuming, in defiance of reason, that he already possesses it. To spin romances is, indeed, man's proper and peculiar function in a world where he only of created beings can make no profitable use of the truth about himself [sic?]. For man alone of animals plays the ape to his dreams. . . . [T]hat the goal remains ambiguous seems but a trivial circumstance to any living creature who knows, he knows not how, that to stay still can be esteemed a virtue only in the dead (p. 50).

Early in his fourth chapter Cabell acknowledges that "no form of greatness is appreciable save in perspective" (p. 87). "Genius", he says, "like Niagara, is thus most majestic from a distance; and indeed, if the flights of genius are immeasurable, its descents are equally fathomless" (p. 88). But this does not materially affect his principal point, which is to juxtapose the life of (attempted) creativity to "the shiftless cult of mediocrity" (p. 86), to the life, honored by precept and proverb, of "a staid and conventional course . . . pursued, upon the indisputable ground that this is the surest avenue to a sufficiency of creature comforts . . ." (p. 87). Yet, he notes, "it has been the deviators from the highway, the strayers in by-paths . . . whom men, led by instinctive wisdom, have elected to commemorate . . . [to sing] the praises of those who have flown in the face of convention, and have notoriously violated every rule for securing an epitaph in which they might take reasonable pride" (p. 87). These people have rejected the virtue of "being practical" (p. 106) in the sense of conforming to the conventionally sensible (pp. 112–113). Those who have viewed life sensibly, those who "live temperately, display edifying virtues, put money in bank, rise at need to heroism and abnegation, serve on committees, dispense a rational benevolence in which there is in reality [he acknowledges] something divine, discourse very wisely over flat-topped desks, and eventually die to the honest regret of their associates" (p. 112), these people, however, leave nothing of "enduring increment . . . nothing durable to signalize his stay upon this planet" (p. 113).[5] Cabal invokes Marcus Aurelius as affirming the view that "by making any orthodox use of your body and brain, you can get out of them only ephemeral results. For all this code of common-sense, and this belief in the value of doing 'practical' things, would seem to be but another dynamic illusion" (p. 114).

Although Cabal is self-consciously dealing with literary creativity – and uses "creative" as a primitive term, whereas that which is and is not deemed creative is ultimately subjective – the professional economist will be inclined to extend the population through the concept of entrepreneurship. Defining the concept as a specific person or group of persons, the case can be made that Cabell's conception of innovative, even iconoclastic, creativity applies to them. And defining the concept in terms of a specific function, a comparable case can be made that Cabell's conception applies to all economic actors, to all persons. Still, however useful such extensions may be, the analyst must be cautioned against trivializing Cabell's reasoning. That this is warranted is particularly evident when one considers Cabell's third step.

But first, Cabell's theory of creativity may suggest but is not, or not very close to, Friedrich von Wieser's, Frank William Taussig's and Joseph A. Schumpeter's theory of leadership. Still, the point is the same: The future of a society very much depends on the respective roles and importance – as well as the specific fields in which, and direction which, Cabellian creativity operates – of the two types of person, the creative and the prosaic, in the economy which together they comprise.

Third, Cabell transforms his analysis of the creative and prosaic lives into an analysis of the abstract, or the withdrawn, and the practical. Cabell argues that "being practical" is fallacious (p. 106; elsewhere he writes of the "vital falsity of 'being true to life' " (p. 118)). His focus is on permanence. He lauds the writer who is able "to make something that may, with favoring luck, be permanent" whereby the author "perpetuates his dreams", whereas the writing of the noncreative author "is certain very soon to require revision into conformity with altered conditions, and is doomed ultimately to interest nobody" (p. 115). (Here the extension to entrepreneurship somewhat breaks down, inasmuch as the entrepreneur(ial function) must adjust to changing conditions.)

But the creative writer does not live in a vacuum. Cabell argues that "the elect artist, who is above all else an economist" (p. 116), must engage in a series of tradeoffs.

> So always this problem [he writes] confronts the creative writer, as to what compromise is permissible between his existence as an artist and his existence as an ephemeral animal (p. 117).

The creative author seeks to produce "enduring literature" (p. 117), but he or she likely must sell their product in order to have income. Presumably Cabell learned this lesson through his own life, early and late, the hard way. (In 1911, unable to find a market for his writings, Cabell tried his hand as a coal-mine operator in West Virginia. The experiment lasted two years. (Wells 1981, p. 108) In other respects, notably what he has to say below about withdrawal, his ideas must have been either learned or reenforced by experience.)

The choices both as to lifestyle and within lifestyle (the compromises noted above) are presented by Cabell in manifestly economic terms:

> the elect artist voluntarily purchases loneliness by a withdrawal from the plane of common life, since only in such isolation can he create. No doubt he takes with him his memories of things observed and things endured, which later may be utilized to lend plausibility and corrobative detail; but, precisely as in the *Book of Genesis*, here too the creator must begin *in vacuo*. And moreover, he must withdraw, for literary evaluation, to an attitude which is frankly abnormal. The viewpoint of 'the man in the street' is really not the viewpoint of fine literature; . . . It is thus from his own normal viewpoint that the artist must withdraw . . . [ellipsis in original] And sometimes the mind goes of its own accord into this withdrawal, and reverie abstracts the creative writer from the ties and aspirations of his existence as a tax-payer. . . . Then it would seem that this ruthlessly far-seeing economist induces such withdrawal by extraneous means (as people loosely say) as a matter of course, and by mere extension of the principle on which he closes his library door. . . . For now he is conscious of stupendous notions; he comprehends the importance of writing down these notions as he alone can write them; and feeling himself to be a god, with eternity held in fee, he need not grudge the slow and comminuted labor of getting all his lovely words just right. And now he is for the while released from inhibitions which compel him ordinarily to affect agreement with the quaint irrationalities of 'practical' persons. For in his sober senses, of course, the economist dare not ever be entirely himself, but must pretend to be, like everybody else, admiringly respectful of bankers and archbishops and brigadier-generals and presidents, as the highest developed forms of humanity[6] (pp. 118–120).

It is in this connection that Cabell raises, indeed invokes, the use of drugs and alcohol as the midwife of literary creativity.

So it is from his own double-dealing that he induces a withdrawal; and with drugs or alcohol unlocks the cell wherein his cowardice ordinarily imprisons his actual self. Nor with him does there appear to be any question of self-sacrifice or self-injury, since, as he can perceive with unmerciful clearness, a man's brain and body are no more a part of him than is the brandy or the opium. All are extraneous things; and are implements of which the economist makes use to serve his end. So the abstraction is induced, the dream is captured

(p. 120).Cabell is sincerely serious about this. But his emphasis on the putative utility of drugs or alcohol should neither obscure nor negate his basic economic argument. Cabell is both juxtaposing two modes of life and indicating the opportunities for marginal compromise which must regularly be faced. A person is tempted to believe that

> It is not the part of a well-balanced person . . . to think of such "economy", nor to appraise a man's relative importance in human life, far less in the material universe, after any such high-flown and morbid fashion, so long as there is the daily paper with all the local news. So we take refuge in that dynamic illusion known as common-sense; and wax sagacious over state elections and the children's progress at school and the misdemeanors of the cook, and other trivialities which accident places so near the eye they seem large. . . . They seem to have had the root of the matter (pp. 121–122).

But

> perhaps the creative writer will continue indefinitely to abuse and wreck that inadequate human body which is his sole medium of expression, in an endeavor to compel the thing to serve his desire (pp. 122–123).

Interestingly, the compromises are worked out both deliberatively and non-deliberatively:

> It may be, of course, that he also is sometimes led by instinctive wisdom, and achieves economy with no more forethought than bees devote to the blending of honey: even when the case stands thus, the fact is in no way altered that actually the creative writer, alone of mankind, does in a logical fashion attempt the unhuman virtue of economy. Whether consciously or no, he labors to perpetuate something of himself in the one sphere of which he is certain, and strives in the only way unbarred to create against the last reach of futurity that which was not anywhere before he made it. . . . [that] it will be his book alone that will endure (p. 123).

Such a person will be considered odd and impractical. Yet in time it will not be he who is condemned as "the reputed wastrel who played the usurer with his loaned body, and thriftily extorted interest, while those contemporaries who listened to the siren voice of common-sense were passing in limousines toward oblivion" (pp. 123–124). Still, Cabell grants that as between the verbal artist and the practical person, "when it comes to deciding which is in reality the wastrel, there seems a great deal to be said for both sides" (p. 124).

Here, then, is the ultimate economy of James Branch Cabell. It involves every author's chosen orientation toward the ordinary world of day-to-day affairs, both in the large and in the details, and all the compromises that all that entails. Every self-reflective research scholar, not only every writer of fiction, should be able to resonate with Cabell's economy. (For such a person, Cabell's argument may not even be novel.) Part of his argument, after all, is that the practical follower of common-sense is unaware both that such an economy exists and that they have made, willy nilly, their own choices in it.

* * * * *

James Branch Cabell's analysis of the economy of Christopher Marlowe – and, by implication, other creative persons – must be of significance for the economist seeking deeper and less mechanical meaning. Cabell's analyses demonstrate the heuristic utility of the tool of the concept of the economic aspect of things. But his analyses also demonstrate that the economic-aspect approach need not be limited to and conducted solely within the Neoclassical/ Chicago School research protocol, with its focus on unique determinate optimal equilibrium solutions deployed within a harmonistic normative paradigm and *Zeitgeist*. More is arguably involved in the economic aspect of life than narrow and mechanical individual rational calculation, constrained maximization, the weighing and comparing of benefits and costs, and so on. The economic-aspect approach can be pursued in conjunction with a program of studying the process by and through which things are worked out, the forces at work, and the various actual workings out. The approach can cope with the opportunities, the tragedies, and the implications for life-style choices, commitments, and conversions.

And more is involved than the role of drugs and alcohol, however much it may have played the midwifery role, or the role of psychic balm. Cabell's analyses point to the struggle to be creative, intellectually or otherwise, a struggle involving the necessity to cope with and endure the differences between oneself and those who lead more "practical" lives – and a struggle, too, within oneself between the creative and the practical courses. What is involved is deeper and more vital than choices as to marginal adjustments. What is involved are choices as to ways of life. More is involved than choices between books and yachts. And the economic aspect of all things pertains to all persons, entrepreneurs, as a category of the creative, included; all persons, each in his or her own way, as each fashions an identity and mode of living.

But Cabell's analysis should not have ended where it did. Should one assume that all creativity is apriori constructive and desirable? "Creativity" is neither a self-defining nor a conceptually self-subsistent category. If, or to the extent that, the iconoclastic creative person is driven by an instinct for domination, such creativity may be arguably misguided. Even absent an instinct for domination, the substantive results of creativity are subject to normative and/or utilitarian/pragmatic evaluation. Blind ambition may engender creativity that encompasses putative evil. The society dominated by a manipulative elite – deemed creative by their own standards – described by Nicolo Machiavelli and Vilfredo Pareto, is not very attractive, especially as regards the negative and inhumane treatment which ruling (and pretender) elites inflict on people effectively deemed "inferior" or "unimportant". On the one hand, the point of these thinkers is that society, polity and economy are in fact systems of mutual manipulation by ambitious, dominance-oriented, self-defined creative people. On the other hand, the affirmation of the creative over the prosaic life is already normative, and the further normative question necessarily arises as to which directions creativity can take are to be preferred and which prevented, and how. In the context of both the economy of James Branch Cabell and the economy of the professional economists, the social construction of social change and social control, and thereby of creativity, is at the heart of the processes of valuation and of working things out.

One further aspect of Cabell's discussion is his recognition of the principle of unintended and/or unforeseen consequences. He finds "perturbing . . . that depravity may, in the last quarter of every other blue moon, be positively praiseworthy" (p. 103) and that "reasoning very often conducts one to undesirable results, and after all has no claim to be considered infallible" (p. 121). The former statement brings to mind Bernard Mandeville's argument, in his *Fable of the Bees*, how practices deemed vice by one moral code may nonetheless lead to social benefits, and comparable lines of reasoning in Adam Smith's *Wealth of Nations*. Both statements serve to indicate that consequences need to be judged and that intended and/or unforeseen consequences are not to be denigrated on that account alone, but too must be judged. What is worked out are the moral and legal rules by which some consequence is deemed vice or virtue. And as I have already quoted Cabell in another connection, in conflicts over such matters, "there seems a great deal to be said for both sides".

Finally, along a different line, if we define practical as the normal practice of economics (normal science in the sense of Thomas S. Kuhn), then the articulation by Cabell of his theory of economy is itself an act of creativity in relation thereto. No mean feat to be reckoned with.

NOTES

1. Although one is tempted to think that the use of "the economy" is metaphorical, it is quite likely that Cabell, and perhaps Freud and Peirce, meant it literally to signify a domain of choice under conditions of scarcity. The scarcity here is neither of resources (for Alpha to use a resource means than Beta cannot, or that for Alpha to put a resource to one use precludes another use by Alpha) nor of rights (for Alpha to have a right

means Beta has a non-right, if both are in the same field of action), but of identity (the choice between, or balancing of, the creative and the prosaic) and of lives (the choice between, or balancing of, the withdrawn and the practical) (see below).

2. Alas, I cannot recall who, how or when someone, several years ago, called this work to my attention; someone did, and to him or her I am both grateful and apologetic. All references in the text not otherwise identified are to this book.

3. I have relied heavily on Wells 1981 and the entries on Cabell in *Webster's Biographical Dictionary* and Locher 1982.

4. Among other types of psychic states, Vilfredo Pareto identified instincts for combination and therefor change, and activity (self-realization), the combination of which may come close to Cabell's creative person. Numerous cultural, psychological and sociological taxonomies exist. What is important is not Cabell's specific taxonomy but his conception of economy formed thereby.

5. The historian of economic thought, upon reading the last phrase, likely will recall to mind Adam Smith's concept of productive labor as involving, among other things (e.g. being tangible and vendible), something durable.

6. This language evokes memories of the phraseology used by Thorstein Veblen. Cabell's distinction between the prosaic/conventional and the withdrawn modes of living seem readily to apply to Veblen the intellectual; but does not parallel Veblen's teleology-matter of fact dichotomy; the two run along quite different axes.

REFERENCES

Cabal, J. B. (1919). *Beyond Life*. New York: Robert M. McBride & Company.

Locher, F. C. (Ed.) (1982). *Contemporary Authors*, vol. 105, Detroit, MI: Gale Research, p. 100.

Wells. A. R. (1981). James Branch Cabell. In: J. J. Martine (Ed.), *American Novelists, 1910–1945*, Part I; Volume Nine of *Dictionary of Literary Biography* (pp. 107–117). Detroit, MI: Gale Research.

Webster's Biographical Dictionary (1971). Springfield, MA: G. & E. Merriam Co., p. 225.

AMERICAN INSTITUTIONALISM

WESLEY CLAIR MITCHELL'S
MONEY ECONOMY AND ECONOMIC EFFICIENCY

Jeff Biddle and Luca Fiorito

INTRODUCTION

The piece entitled "Money Economy and Economic Efficiency," which follows this introduction, was transcribed from a typescript found in the Wesley Clair Mitchell papers in the Butler Library of Columbia University. On the first page of the typescript is the pencilled notation "1924–25"; it is not known who wrote these dates, but as will be argued below they almost certainly do not identify the time at which Mitchell originally wrote the material in the typescript. There are, however, bases for speculating about both when the material was written, and the purpose for which the typescript was prepared.

Mitchell's doctoral dissertation and early post-doctoral research was concerned with the economic impact of the issuance of unbacked paper currency or "greenbacks" during the American civil war. As Mitchell finished up his second book on the greenbacks around 1906 (Mitchell, 1908), however, he began to plan a more ambitious research project, which he came to call "the Money Economy". In modern economies, Mitchell noted, economic activity took the form of making and spending money. People participated in production in order to make money, and spent the money so earned on consumption goods produced by others. This separation of production from consumption was made possible by a "highly organized group of pecuniary institutions", including "an imposing complex of inter-related, everchanging

Research in the History of Economic Thought and Methodology, Volume 19A,
pages 77–84.
2001 by Elsevier Science B.V.
ISBN: 0-7623-0703-X

price agreements." Mitchell's planned book on the Money Economy was to be a theoretical account of the operation of these institutions and their influence on men's habits of thought and patterns of behavior. He later expanded his plan for the project to include an account of the origin and evolution of the pecuniary institutions of the money economy. (Mitchell, 2000; Lucy Mitchell, 1953, p. 167, Rutherford, 1996, p. 318.)

By Mitchell's own account, his conceptualization of the Money Economy and the proper approach to studying its operation and impact owed much to Veblen, both directly and through the ethnological reading that Veblen had encouraged him to undertake. Mitchell mentioned Simmel as an influence as well. Through 1907 and 1908 Mitchell wrote several draft chapters of the Money Economy, but considered the chapters to be "unsatisfactory" (Lucy Mitchell, 1953, p. 173). By 1909, he was expressing his disappointment with the overly speculative nature of what he had produced to his future wife:

> I used to say to myself that the Money Economy should not have a statistic between its covers. But that I am beginning to think a foolish idea. It is slovenly to rest content with "very littles," "great deal mores" and "great importances", when even approximate precision of knowledge and statement are attainable. I want to prove things as nearly as may be and proof usually means an appeal to the facts – facts recorded in the best cases in statistical form . . .
>
> This feeling has been growing on me as I have realized how slight an impression Veblen's work has made upon other economists. . . . The fact that his works lack (in appearance far more than in reality) a basis of exact investigation further gives even a feeling that it is ingenious speculation – stimulating, but not to be reckoned with seriously. (L. Mitchell, 1953, p. 176.)

Mitchell's reaction to this uneasiness was to undertake an analysis of economic fluctuations that was grounded solidly in recent statistical data from four countries. The result was the classic *Business Cycles*, published in 1913.[1]

At the time Mitchell began serious research into business cycles, he did not see it as an alternative to the Money Economy project. Instead, he saw the careful treatment of business cycles as an introduction to a more wide ranging theory of the origins and workings of the money economy. However, constructing an empirically grounded theory of the business cycle itself became a project that occupied Mitchell for the rest of his life. It was the main focus of his research after 1913, and the work for which he became famous. The manuscript chapters of the Money Economy that he had produced prior to 1910 were never published as a book. Mitchell did not abandon them, though. As Dorfman (1949, p. 459) notes, many passages in *Business Cycles* were based on material in the draft chapters, and Mitchell also developed addresses and articles out of that material, including "The Backward Art of Spending Money

(1912)," "Human Behavior and Economics (1914)" and "Making Goods and Making Money ([1923] 1937)."[2]

It is our opinion that the essay "Money Economy and Economic Efficiency" is based on a draft chapter of the unpublished Money Economy project. However, it appears that Mitchell was attempting to shape this material into a journal article: it is written so as to stand alone, rather than building on material previously developed as would a book chapter; and in the fifth paragraph Mitchell speaks of "the purpose of this paper". It includes footnotes, which Rutherford (1996) points out were not typically found in manuscript versions of Mitchell's public addresses. Also, it appears to be unfinished – unlike Mitchell's published articles, the typescript ends without a concluding section or summing up of any sort. We also believe that despite the date penciled on the top, the prose in this typescript was for the most part written prior to 1913, maybe even before 1910, and revised only minimally if at all thereafter. Our reasons for believing this will be laid out as we discuss the content of the essay.

In the essay itself, Mitchell begins by describing the characteristics of what he calls the money economy – the complex division of labor, production for the sake of profit rather than consumption, consumption of goods produced by others, all coordinated by the widespread use of money and a complex system of price bargains. He then lays out the questions to be addressed by the paper "How does the consummate money economy of today which constrains the individual to aim at making money affect the volume of serviceable goods produced and consumed by society?" Mitchell briefly affirms the conventional wisdom that institutions which enlist self interest as a motive of economic activity tend to promote economic efficiency, but moves quickly to the material that constitutes the bulk of the essay: a litany of ways in which modern pecuniary institutions actually inhibit efficiency. To illustrate that the results of production guided by self-interest often clash with the social interest, Mitchell mentions child labor, the tendency to monopoly, waste of natural resources, fraud, and coercion. He asserts that laws to prevent such activities at best remove symptoms of a "chronic disorder" of society. To show that the motive of self-interest is sometimes too weak to promote efficiency in the modern economy, Mitchell points out that stockholders seldom actually run the corporations whose profits they receive, and that for most wage workers there is little relationship between the effort they exert on the job and their monetary reward. Again, existing attempts to remedy these situations are merely "palliatives for a chronic disorder." Mitchell next mentions the waste created by the lack of coordination between firms in an interdependent economy;

unfortunately, modern efforts to increase the coordination between business firms are usually undertaken by businessmen in pursuit of monopoly power.

Mitchell then turns from the volume of goods produced to the types of goods produced in the money economy. Citing Weiser's *Natural Value*, he describes how production guided by price and profit combined with an unequal distribution of wealth leads to a diversion of resources towards production of luxuries for the rich, when devotion of the same resources to producing necessities for the poor would increase the efficiency and welfare of the working class. Lest it be argued that it is inequality rather than the money economy per se that causes this problem, Mitchell points out the various means by which modern economic institutions act to preserve inequality.

Mitchell describes and criticizes the way in which pecuniary institutions determine the identity of "the men to whom the direction of the process of production is entrusted," and with a nod to Veblen he notes that when the purpose of producing goods is to make money, it is experts in money making and not the technical experts who win the competition to direct production. The final pages of the essay describe the ways in which pecuniary institutions have injected a counter-productive uncertainty into economic relations. Those in charge of production must invest and produce based on forecasts of future demand, and wrong guesses lead to waste. "Such mistakes," writes Mitchell "are not accidents, they are certainties produced by the economic organization." People must make their economic plans and decisions based on market prices, but these prices are constantly changing in an unpredictable manner, not only because of the forces of supply and demand, but because of changes in the money supply brought about by gold discoveries or the machinations of government. At this point Mitchell comes close to the topic that would dominate his later work. Citing a 1900 book by E. D. Jones, he describes how shared feelings of optimism or pessimism on the part of producers lead to "business crises and periods of depression," and asserts that constant and confusing price changes sometimes stimulate and sometimes depress the rate of production. With this argument, the typescript ends.

Two overall characteristics of the essay would seem to mark it as a product of Mitchell's pre–1913 period rather than post 1920. First, there is the fairly prominent presence of Veblenian themes, particularly the importance of the dichotomy between making goods and making money, that also marks other recently published material from the Money Economy project (Mitchell 1995, 1996). Second, the essay is perhaps more strongly and consistently critical of the modern economic system than anything Mitchell would publish after 1913. It is instructive to compare the essay to sections of *Business Cycles* that appear to have been based on the same "Money Economy" source material. The opening paragraphs of chapter II of Business Cycles, titled "the Economic Organization Today" begins with a description of the Money Economy similar to that in the early paragraphs of the typescript. A later section in the chapter, "The Alleged Planlessness of Production," refers to the contrast between the coordination within business enterprises and the lack of coordination between them. Mitchell then writes:

> This union between encouragement of individual efficiency and opportunity for wide cooperation is a great merit of the money economy. It provides a basis for what is unquestionably the best system of directing economic activity which men have yet practiced. Nevertheless, the system has serious limitations . . . (Mitchell, 1913, p. 38).

The question of whether the money economy encourages the right balance between individual initiative and cooperation is also one of the main themes of "Money Economy and Economic Efficiency," but Mitchell's affirmation of the value of the present system is much stronger in *Business Cycles* than in the essay.

Mitchell continues in *Business Cycles* by listing four problems (pp. 38–40). The first is the lack of coordination between businesses. This is discussed in point 3 of the essay, and the second paragraph of this discussion appears almost verbatim in *Business Cycles*. However, some of the of the negative consequences of this problem covered in the essay, such as the timing of combinations for business rather than technical purposes, are not in *Business Cycles*. Indeed, Mitchell mentions the increasing size of business only as a positive phenomenon without mentioning the danger of monopoly. The second problem covered in *Business Cycles* is that managerial skill is devoted to making money rather than producing goods. Here one finds only a brief mention of what Mitchell is very concerned with in the essay – that inequality leads to irrationality in the mix of goods produced in the money economy. The third problem is the impact of uncertainty on the planning efforts of businessmen, a matter covered more extensively in points 8 and 9 of the essay. The fourth problem is the role of lenders in controlling the production process. In an earlier section of *Business Cycles* (p. 34–35) Mitchell had argued that these lenders are not always qualified to make intelligent decisions about the direction of economic activity. The complaint of the essay that admission to the class of lenders is denied to potentially capable scions of the lower classes, however, does not appear in *Business Cycles*.

Overall, then, Mitchell's discussion of the problems of the money economy in *Business Cycles* was less comprehensive, much less detailed, and more carefully qualified than that of "Money Economy and Economic Efficiency." This would be the case as well in Mitchell's later published work dealing with the problems of the money economy. In "Making Goods and Making Money",

Mitchell spoke to an audience of engineers of the "grave consequences" of "subordination of our common interest in making goods to our individual interests in making money," but before listing them, he "emphatically stated" that "the money economy is doubtless the best form of economic organization for promoting the common welfare that men have yet devised" (Mitchell [1923] 1937, p. 144). He then describes problems related to an unstable dollar and to business cycles. In "Intelligence and the Guidance of Economic Evolution", an address given during the depression, Mitchell offers a historical outline of "the difficulties encountered as the system of free enterprise unfolded" – child labor, fraud, exploitation of workers and consumers by businessmen, cutthroat competition, class antagonism, a tendency to monopoly, Veblen's "business sabotage", inequality, and business cycles. But Mitchell also reminds his audience that the money economy has brought technological improvements, rising living standards, and declining death rates (Mitchell [1936] 1937, pp. 118–121). Similar passages can be found in other Mitchell addresses and articles, e.g. "Economic Resources and Their Employment (Mitchell 1941, p. 7–13); and "The Economic Basis for Social Progress (Mitchell 1931, pp. 47–49). In these addresses, Mitchell lists some of the same problems described "Money Economy and Economic Efficiency," but the tone in these later addresses is different. "Money Economy and Economic Efficiency" presents a picture of an economic system that is seriously and fundamentally flawed – the problems of the money economy are presented as "chronic disorders" of modern economic organization; there is a sense that they are contradictions inherent to the system, so that attempts at reform only remove symptoms, or act as palliatives. In the later work, descriptions of problems of the money economy come as part of a call to reform an economic system that basically has much to recommend it, although a central theme of Mitchell's message is that future attempts at reforms must be based on sound scientific knowledge of how that system works.[3]

Perhaps the most compelling reason for believing that "Money Economy and Economic Efficiency" was not written in the 1920s is its treatment of business fluctuations. After 1913, in almost any of Mitchell's addresses dealing with the ills of the modern economic system, the problem of business cycles was given a place of prominence, and understandably so – Mitchell was known as an expert on business cycles, and often he was attempting to make the case that research into business cycles of the sort that he conducted at the National Bureau was worthy of support. In "Money Economy and Economic Efficiency," however, the phrase business cycles never even appears. Also, when Mitchell does discuss "business crises and periods of depression", the only theory of crises referred to comes from a 1900 book by E. D. Jones.

However, in the course of writing *Business Cycles* between 1910 and 1913, Mitchell became an expert on the literature of crises, depressions and cycles. The 1913 book had an entire chapter devoted to reviewing and classifying "current theories concerning the business cycle." The Jones book was mentioned only in a footnote, in a section dealing with "early theories of crises" (Mitchell 1913, Chapt. 1; p. 3). If "Money Economy and Economic Efficiency" had been written or even substantially revised in the 1920s, it certainly would not have contained such an abbreviated and primitive discussion of the place of business cycles in the money economy.[4]

Indeed, a good measure of how Mitchell's style of research and presentation changed between the time he was writing "the Money Economy" and the 1920s can be seen by comparing Chapter II of *Business Cycles*, which Dorfman (1949) argued was based largely on the unpublished "Money Economy" chapters, to Chapter II of the 1927 book *Business Cycles: The Problem and its Setting*. The latter volume was the first installment of Mitchell's attempt to expand and improve the 1913 book, and was designed to have a parallel structure. Although the corresponding chapters in the two books share many of the same section headings, the assertions and arguments in the 1927 volume are illustrated with evidence from statistical studies or supported by with references to tables and figures, recalling Mitchell's desire, as expressed to his wife in 1909 as he put aside the Money Economy project, to "prove things as nearly as may be" by appealing to facts and statistics.

What one sees in the following essay, then, is an example of the early thought of Wesley Mitchell – more reflective of a Veblenian influence and less empirically grounded than his later work. It is evidence that the young Mitchell was, if not more critical of capitalism, more willing to criticize capitalism publicly than the older Mitchell. On the whole, the essay offers interesting evidence on the development of the thought of one of the most prominent and influential economists of the 20th century.

NOTES

1. Years later in 1928, in a written response to a query by J. M. Clark, Mitchell would again attribute the change in his research focus around 1909 to a concern with the overly speculative nature of the money economy project, and would repeat his criticisms of the style, if not the substance, of Veblen's work (Burns, 1952, p. 95, 97).

2. Two of Mitchell's addresses based on Money Economy material have recently been published (Mitchell, 1995, 1996), and Rutherford's (1995, 1996) introductions to those addresses discuss at more length the relationship between the Money Economy manuscript and Mitchell's later published work.

3. Biddle (1998) contains a discussion of Mitchell's beliefs about social reform.

4. Also, all of the references in the typescript are from the first decade of the 20th century or before, and Mitchell took pains to stay current in his reading.

REFERENCES

Biddle, J. (1998). Social Science and the Making of Social Policy: Wesley Mitchell's Vision. In: M Rutherford (Ed.), *The Economic Mind in America: Essays in the History of American Economics*. London: Routledge.

Burns, A. (Ed.) (1952). *Wesley Clair Mitchell: The Economic Scientist*. New York: National Bureau of Economic Research.

Dorfman, J. (1949). *The Economic Mind in American Civilization*, vol. 3. New York: Viking Press.

Jones, E. D. (1900). *Economic Crises*. New York: Macmillan.

Mitchell, L. S. (1953). *Two Lives: The Story of Wesley Clair Mitchell and Myself*. New York: Simon and Schuster.

Mitchell, W. (1908). *Gold, Prices, and Wages under the Greenback Standard*. Berkeley, CA: The University of California Press.

Mitchell, W. (1912). The Backward Art of Spending Money. *American Economic Review*, 2, 269–281.

Mitchell, W. (1913). *Business Cycles*. Berkeley, CA: U. of California Press.

Mitchell, W. (1914). Human Behavior and Economics: A Survey of Recent Literature. *Quarterly Journal of Economics*, 29, 1–47.

Mitchell, W. ([1923] 1937). Making Goods and Making Money. In: *The Backward Art of Spending Money* (pp. 137–148). New York: McGraw-Hill.

Mitchell, W. (1931). The Economic Basis for Social Progress. In: *Proceedings of the National Conference of Social Work, Fifty-seventh Annual Meeting, June 1930* (pp. 34–49). Chicago, University of Chicago Press.

Mitchell, W. ([1936] 1937). Intelligence and the Guidance of Economic Evolution. In: *The Backward Art of Spending Money* (pp. 103–136). New York: McGraw-Hill.

Mitchell, W. (1941). Economic Resources and their Employment. In: *Studies in Economics and Industrial Relations* (pp. 1–23). Philadelphia: University of Pennsylvania Press.

Mitchell, W. (1995). The Criticism of Modern Civilization. *Journal of Economic Issues*, 29, 663–682.

Mitchell, W. (1996). Money Economy and Modern Civilization *History of Political Economy*, 28, 329–357.

Mitchell, W. (2000). Money Economy and Economic Efficiency. *Review of the History of Economic Thought and Methodology* (this volume).

Rutherford, M. (1995). Introduction to 'The Criticism of Modern Civilization' by Wesely Clair Mitchell. *Journal of Economic Issues*, 29, 658–662.

Rutherford, M. (1996). An Introduction to 'Money Economy and Modern Civilization' by Wesley Clair Mitchell. *History of Political Economy*, 28, 317–328.

Wieser, F. (1893). *Natural Value*. London: Macmillan & Co.

WESLEY CLAIR MITCHELL'S *MONEY ECONOMY AND ECONOMIC EFFICIENCY*

Edited by Luca Fiorito

Economic life among Western Nations is characterized by an unprecedentedly wide use of money. Production is carried on by business establishments for the sake of money profits, and is guided by pecuniary considerations. Distribution depends on money wages, money interest, money rents, and money profits. Consumption requires money income with which to buy goods in markets at money prices. To be sure, families still produce some of the goods and still perform some of the services they require; they frequently raise their own vegetables and generally cook their own food. But for a long time this direct method of procuring goods has been declining, and it is now confined within narrower limits than ever before. The economically important thing for a family is now its money income, and to get a money income its members must take such advantage as they can of the pecuniary opportunities afforded by their social environment. Economic efficiency means to them not ability to produce serviceable goods, but ability to make money. The whole economic process in which they share is carried on by an imposing complex of inter-related, ever-changing price-agreements made between individuals or groups of individuals.

Contrasting strongly with these patent facts of the money economy, stands the equally patent fact that men are animals which must have food to eat,

Research in the History of Economic Thought and Methodology, Volume 19A, pages 85–102.
2001 by Elsevier Science B.V.
ISBN: 0-7623-0703-X

clothing to wear, and roofs to shelter their heads. The elaborate pecuniary organization of modern life does not emancipate the men who live it from dependence on such material goods. Economic welfare is not a matter of the pecuniary profitableness of business enterprise, but a matter of the relative abundance of the material goods produced and consumed.

The distinction between making money and producing material goods is not of much practical importance to the man on the street, because he can usually get whatever material goods he may desire provided he has money enough. The difficult thing is to make the money. He devotes his attention mainly to this problem, and seldom thinks much about material goods except in pecuniary terms. This attitude of mind is prevalent because making money is for the individual equivalent to procuring material goods.

The economists' attitude toward making money and producing material goods, is just the reverse of that of the man on the street. He puts much stress on efficient production and little stress on making money. The reason is that the economist is concerned with the economic life of large communities, not of single individuals. A nation's supply of material goods does not depend on the amount of money its citizens make, but on the physical quantity of the goods they produce at home and acquire from abroad. The economist accordingly devotes his attention to the way in which men produce, evaluate, and distribute these material goods. He treats money as a piece of mechanism employed to facilitate these processes, important in that it enhances the amount of these goods which a community can produce. He is inclined to rebuke the man on the street for laying so much stress on money-making. But from the economist's own point of view, that quantity of material goods is the matter of importance, the man on the street is correct. For the individual, tho not for the nation, to make money is the way to get material goods.

So much, indeed, is the economist preoccupied with what he regards as the deeper facts of economic life that in his general theory he pays but scant attention to money and prices. To him the economic process is substantially the same whether money is used or not. It is therefore well to put aside the superficial, but complicated, facts of money and prices in order to get an unobstructed view of the philosophical realities which underlie them. Accordingly, he writes of men in a money economy in terms more appropriate to man in a natural economy. He practically imputes his own interest in material goods to individuals who are really interested in making money. But after discussing his "fundamental problems" in this fashion, he frames a separate theory of money, and warns the reader that the two disparate discussions must be pieced together to secure a full understanding of modern economic life. By this procedure, the economist gains simplicity; but he sacrifices realism and neglects the influence which the use of money exercise on modern life.

The purpose of the present paper is to discuss one phase of this neglected problem – viz. the effect which the present pecuniary form of economic organization has upon economic efficiency. It deals with the two aspects of modern economic life which stand in such sharp contrast to each other, the pecuniary and the material, and tries to work out the relations between them. More specifically, the problem is: How does the consummate money economy of today which constrains the individual to aim at making money, affect the volume of serviceable goods produced and consumed by society?

I

Observation of communities where economic life is carried on by making and spending money shows the prevalence of this curious situation – men commonly regard themselves as aiming to provide goods for the use of their own families, but they spend their working hours in helping to produce goods which are consumed by others. The economic organization of such communities leaves the individual nominally free to do what work he chooses but it also imposes on him the responsibility of providing for himself and his dependents. By this means it holds out to the individual economic self-interest as a motive for efficient economic activity. But at the same time, the economic organization enables the individual to promote his own interest best by working for others, and thus makes organized cooperation the prevailing economic practice.

The union of self interest and social service in the economic life of the individual appears on a larger scale in the life of society under the guise of a union of individualistic and socialistic institutions. Production as carried on by a thorough going use of money is in external form an eminently socialistic process. It is conducted by interdependent groups of interdependent individuals. Each group gets its materials and apparatus from other groups, and offers its products for the use of the whole community. Within each group the individual members are dependent first on each other's cooperation for the chance to continue producing, second on continued production by other groups for the goods they consume. The whole organization of producers, in fact, suggests an industrial army, maintained by a socialistic state, for the mutual benefit of all its members. But if the form of this organization is socialistic, its animating spirit is individualistic. Men are supposed to enlist in the industrial army only when they can promote their individual interests best by working with others and for others; they make their own terms, and they are nominally free to change from one company to another, or to withdraw altogether if they choose. The captains of industry assume their rank at their own risk, they must hire their subordinates as best they can; their relations with each other are largely hostile, and they plain their campaigns to secure, no public welfare, but private profits.

This union of self-interest and social service in the economic life of the individuals, and of individualistic and socialistic institutions in the economic life of society, rests on the use of money as a means of conducting exchanges. An individual cannot provide for his family by producing goods for other men unless he can exchange his services for the goods his family requires. While exchanges can be made on a small scale by barter, the use of money is indispensable, so far as we know, for exchanges so elaborate as those made by the modern man. Similarly, when we try to imagine society dispensing with the use of money in price-agreements made between individuals, we cannot see how the present régime, characterized by the union of individualistic and socialistic institutions, could be maintained. The economic freedom and responsibility of the individual may be maintained without exchanges based on money, but every man would have to produce most of the goods used by his family – organized cooperation on a large scale would cease. Or organized cooperation on a large scale might conceivably be maintained without voluntary exchanges between individuals by the aid of money; but some authority other than the individual would have to determine what place each man should occupy in the industrial army and what reward he should receive, and the economic freedom and responsibility of the individual would cease. Imagining what would happen under non-existent conditions, however, is less convincing than observation of what is. Anyone who will observe the economic life of the present will find that exchanges at money-prices are the means by which men make work for others consonant with pursuit of their individual interest.

It is by uniting self-interest and social service in the life of the individual, and individualistic and socialistic institutions in the life of society, that the use of money exercises its profound influence on economic efficiency. To investigate this influence accordingly we have to inquire how each of these elements affects efficiency, how far each is allowed free scope for development, and how far the union effected between them by the use of money is harmonious.

The first question can be answered very briefly. Without assuming that all or even most economic activity is due to the individual's conscious consideration of his economic interests, we are nevertheless convinced that a form of economic organization which enlists self-interest among the motives to economic activity promotes economic efficiency. The energy of the modern workman and the enterprise of the modern employer depend in considerable measure on the degree in which they find their economic welfare as individuals affected by the vigor and the discretion of their work. There is even less doubt that organized cooperation among men in production is conducive to efficiency. For all the factors which we commonly think of as the most conspicuous causes of the present economic efficiency are either forms of organized cooperation – e.g. division of labor in all its varieties – or are dependent on such cooperation – e.g. the use of machinery.

It would be interesting to inquire from another point of view how the use of money enhances efficiency, by tracing the gradual development of the present form of economic organization with its ubiquitous use money out of simpler forms in which money was little used. It could be shown that the economic freedom and responsibility of the individual and therefore self-interest as an economic motive, and organized cooperation as an economic practice, have developed together, and that their development has been closely dependent on the increasing use of money. It could be shown also that this development has been accompanied by increasing economic efficiency – that, in the usual evolutionary interpretation of the phrase, this increasing efficiency is the cause of the development. Bur such a discussion to be effective would have to be detailed and a detailed historical discussion is out of the question here.

The other questions – how far the use of money allows scope for the full development of each of the efficiency-producing factors, and how far it brings them into harmony – cannot be answered so briefly. There seems to be an inherent antagonism between the unmitigated sway of self-interest as an economic motive and the unmitigated practice of producing goods for others to consume. The use of money certainly does not remove this antagonism; it merely effects a compromise, successful at some points unsuccessful at others. Wherever the use of money allows the motive of self-interest to interfere with the practice of cooperation, or the practice of cooperation to lessen the effectiveness of a motive of self-interest, there efficiency suffers. That these points of conflict are neither few nor unimportant is shown in the next section.

II

(1) The man in the street commonly speaks of the purpose of economic activity as "making money." This phrase states accurately one result of our peculiar form of economic organization – the individual's conscious aim is in truth to make money. Productive efficiency under this organization depends on the fact that to make money the individual as a rule performs some service for his

fellow men for which they are ready to pay him. But it is notorious that the economic organization offers many opportunities to individuals for making money by acts which detract from instead of increasing the volume of goods produced by the community. For example: in many industries it is possible to make money by employing child-labor, though the productive efficiency of society suffers from the stunted development of the children; the rapid cutting of forests in the United States has brought fortunes to lumbermen, though it is conceded to be an economic injury to the nation; the formation of a monopoly sometimes enables its managers to increase profits while reducing the quantity of goods supplied to consumers; the wrecking of a corporation by those in control is sometimes more profitable to them than its management as a productive enterprise, etc.

Our money-using economic practice is based on the theory that no man will sell or buy except to his own advantage; but the economic organization often permits certain individuals to cajole or compel others to agree to hurtful bargains. And if individuals do not always succeed in defending themselves against deception or superior business strength, still less does society succeed in defending its common interests against men who can make money by invading them. True, society attempts to check by legislation the grosser forms of fraud and oppression of one individual by another, and also to check the social inimical bargains by which both of the parties concerned secure an immediate gain. But such laws accomplish at best nothing but the removal of certain symptoms of a chronic disorder from which society suffers – the fact that under the present form of economic organization men can make money in ways that detract from, as well in ways that contribute to, social efficiency.

(2) In the class of cases referred to there is no lack of individual efficiency. The employer of child-labor may be both skilful and energetic, the wrecker of corporations may lead the most strenuous of lives. The difficulty lies in the economic organization which allows their energy in making money for themselves to assume forms injurious to social efficiency.

It is curious that the same economic organization which allows the pursuit of individual gain to become so intense as to sacrifice social efficiency, in these cases, should in other cases fail by not giving individuals a sufficiently intense motive for exerting their energy in production. But such is the truth and the more elaborate the organization of productive enterprise becomes, the graver is this defect. In a large corporation interest in efficient production is diffused among the stockholders. This location of the interest is not favorable to the highest efficiency because the stockholders are not the people on whom efficiency chiefly depends. Even the directors whom they choose from among their number to manage the enterprise exercise as a rule only a general superintendence. The salaried officials entrusted with direct conduct of the business have not so vital an interest in its success as a man working on his own account – even though their incomes depend in part on the dividends.

More serious is the fact that the great mass of the employees have still less interest than the responsible officials in doing the best they can. A man who is distinctly below the average in efficiency may be discharged, and a man who is distinctly above the average may receive steady employment and promotion. But the dependence of wage-earners' money incomes on their efficiency is not sufficiently close and certain to stimulate everyone to his best efforts.

The gravity of this defect in economic organizations is proved by the many attempts made to remedy it. These attempts usually come from the proprietors who have a direct pecuniary interest in efficiency. The special inducements which some corporations hold out for the purchase of stock by their employees, profit-sharing schemes, premium and bonus systems of paying wages, payment by the piece instead of by time, etc., are primarily intended to promote efficiency of labor by giving the laborer a greater individual interest in working hard. Such schemes, like laws protecting social efficiency against injury from private pursuit of wealth, are palliatives for a chronic disorder of the economic organization. This organization concentrates interest in efficiency in a single economic class – the recipients of profits – and leaves the rest of the workers with an interest that is too slight and too remote to produce the best results.

(3) Just as the present economic organization fails to make the most of self-interest as a factor promoting individual efficiency, so does it fail to make the most of cooperation in the work of production.

Effective cooperation requires not only division of the labor of producing, but also a corresponding coordination between the different parts created by the division. The work of producing is now divided, first among many business establishments, and then divided again within each establishment. Our economic organization provides for effective coordination between the different parts created by this detailed division within establishments; but the coordination which it provides for the different parts created by the larger division between establishments is strikingly defective.

The two systems of coordination differ in almost all respects. Coordination within an establishment is the result of careful planning by experts; coordination between establishments cannot be said to planned at all; rather it is the unplanned result of the natural selection in a struggle for business survival. Coordination within an establishment has a definite aim – the increase of profits by more perfect utilization of productive energy; coordination between establishments has no definite aim – we can see in it the conflicting plans and aims of the units, but no conscious plan or aim of the system as a whole. Coordination within an establishment depends on a single authority, considering a single interest, and possessing power to carry its plans into effect; coordination between establishments depends on independent authorities, considering different interests, and without power to change existing arrangements unless some authority can persuade or coerce the other to accept its plans. As a result of these differences, coordination within an establishment is characterized by economy of energy; coordination between establishments by waste.

Like the other defects of the economic organization which have been mentioned, the waste resulting from a lack of intelligent general plan for the coordination of productive efforts is shown by the efforts made to remedy it. These efforts do not take the form of attempts on the part of public authority to work out a general scheme for the direction of production, but the form of attempts on the part of private persons to harmonize their conflicting interests. Such attempts – the various forms of business alliances, combinations, and integrations – increase productive efficiency by extending the field over which intelligently planned coordination of effort reigns. But this remedy, like the others spoken of, does not effect a cure. First, the combinations do not cover the whole field. Second, combinations are made, not when technical experts are convinced that waste can be diminished by securing closer coordination, but when business experts become convinced that profits can be increased by combinations. Various factors besides productive efficiency enter into the businessmen's calculations of profit – for example, the opportunities for selling corporation securities. It frequently happens, therefore, that it is not until long after the technical advantages of closer coordination have been apparent that the businessmen in control agree to combine.[1] Third, combination as a remedy for waste in production increases the danger for another chronic disorder of the economic organization – the clashing of individual and social interests. For the more complete is the control attained over any brunch of production by a single business authority, the freer becomes that authority from competitive compulsion to regard the interest of the public. Social interests may be better served by a number of competing producers, despite the waste resulting from lack of coordination, than by an economically operating monopoly.

The effect on competition on productive efficiency deserves notice here because of the intimate relation between competition and the three defects of the economic organization which have been discussed: opportunities for individual profits won in ways detrimental to social efficiency, the imperfect functioning of self-interest as a stimulus for individual efficiency, and the waste arising from imperfect coordination of productive efforts. Competition is the stock remedy for one form of the first defect, viz., monopoly; but it aggravates rather than lessens the dangers from other forms – e.g. the employment of child-labor. It is also regarded as a remedy for the second defect – laxity of individual effort; but when applied without restraints it may reduce efficiency by the opposite extreme – "overdriving." It can exist between business establishments only where the third effect – lack of coordination – is found; but then it is highly favorable to the development of effective coordination *within* the rival establishments. Even in this last respect its relation to social efficiency is not simple, because it fosters certain forms of business activities which do not increase the volume of the product though they are necessary for the maintenance of profits – e.g. advertising. On the other hand, attempts to remedy the third defect make the first or second defect more serious, because they repress competition. If the remedy is business combination society runs greater danger from monopoly, if the remedy is public control over production society runs greater danger from decline in individual energy.

(4) Productive efficiency does not mean the production of the largest possible volume of goods, but the most perfect utilization of available productive energy for meeting the needs of a community. From this point of view the kind of goods produced is as important as the volume. We must, therefore, inquire what effect the existing economic organization has in this direction.

Since the present régime sets up money making as the immediate end of economic activity, the men who direct production are constrained to ask, not what goods are most needed by consumers, but what goods will bring them the largest profits. Their guiding principle is, therefore, not relative social needs, but relative money prices and money costs. To be guided by considerations of social needs instead of considerations of market prices is to be a philanthropist instead of a businessman.

This determination of what goods to produce by producer's forecasts of relative money profits is an objective fact about which there can be no doubt. But whether this method of determination prevents productive energy from being utilized to best advantage for supplying consumers' needs is a problem. The possibility of solving this problem depends on the possibility of finding a criterion of the social importance of the wants of different people which can be used in making comparisons.

Economists usually grade wants according to the intensity of the feeling that accompanies the desires or their gratifications. If one accepts this criterion of importance it seems at first sight easy to show that much productive energy is misdirected by the present method of determining what goods to produce. As Von Wieser puts it, production furnishes "luxuries for the wanton and the glutton, while it is deaf to the wants of the miserable and the poor."[2] But this easy solution of the problem overlooks the fundamental difficulty of measuring and comparing the intensity of the feelings of different people. We have no common denominator which we can apply, and without a common denominator of feeling what logical justification have we for saying even that the intensity with which luxuries are desired by the glutton is less than the intensity with which necessities are desired by the poor?

Fortunately there is a less subjective standard than intensity of feeling by which the social importance of wants can be gauged – the relation of their satisfactions to progress in social welfare. It is true that progress in social welfare is a vague phrase and is interpreted in different senses by different men. But in the Western world, at least, the meaning consciously or unconsciously attached to it by most men is that the member of the community should "attain, generation after generation, a wider and fuller life by developing increased faculties and satisfying more complicate desires." Anyone to whom this interpretation of progress is acceptable cannot regard with complacency the present system of directing production by considerations of market price. For a system which allows an appreciable part of the productive energy of a community to be used in making luxuries for the rich while large number of the poor lack sufficient goods to sustain their productive efficiency unimpaired is not favorable to the development of increased faculties and the growth of more complicated desires among the population as a whole. From this point of view, therefore, productive energy cannot be said to be utilized to the best advantage under the present economic organization. It is misdirected whenever intelligence, labor, and materials are employed in making goods whose consumption contributes less than other goods which might have been made in their stead to progress in social welfare. Such misdirection clearly exists so long as the possibility remains of increasing the efficiency of large classes of the community by devoting a large share of productive energy to making more food, clothing, houses, etc., for their use.

(5) The fact that our economic organization makes price the guide in determining what goods shall be produced would not necessarily result in misdirection of productive energy were all individuals equally able to pay for the goods they need. It is because there are many people who cannot pay remunerative prices for the goods necessary to maintain individual efficiency, and because there are others who can pay remunerative prices for luxuries, that misdirection occurs. The use of money, therefore, appears to be only an innocent intermediary in causing the mischief; the real culprit is inequality of riches.

Recognition of this fact, however, does not exonerate the use of money from blame unless it can be shown that the use of money has no responsibility for this inequality. Now there is no doubt that inequality of riches rests in part on causes with which the use of money has at most a remote connection. Men are born into the world with unlike economic capacities and unlike advantages of environment. Even if every family produced for itself alone and made no use of money, innate differences in strength, dexterity, industry, thrift, and intelligence, as well as differences of physical environment, would lead to inequalities of riches. But it is nevertheless true that the existing inequalities are due in large measure to our peculiar form of economic organization with its all-pervading use of money.

First, by increasing the complexity of economic life the use of money has widened the scope for the exercise of whatever economic capacities men are born with, and has therefore enabled the better endowed or better placed individuals to surpass their less fortunate fellows further in accumulating riches. Just as the merchants in a large city show greater disparity of fortunes than the merchants in a hundred small towns, so the merchants of a nation today show greater disparity of fortunes than the merchants of earlier centuries. In both cases the result is due in large part to the difference in the business opportunities afforded by the highly complex and the relatively simple economic organizations. Second, the existing economic organization affords peculiarly effective means of turning present excess of income over current consumption into a source of future income, and therefore helps in perpetuating, and so increasing, the disparity of riches which exist at any moment. In technical terms, the use of money as investment and loan capital makes disparity of riches cumulative. Third, this disparity is cumulative in another way. The economic organization, by fostering division of labor, has led to the development of elaborate technical methods in almost all departments of economic activity. The man who would enter the more lucrative occupations must secure special training of an industrial, professional, or business sort, and in many cases he must also have at his disposal considerable funds before he can set up for himself. The expensiveness of this training in time or money, the difficulty of securing opportunities for experience, and the difficulty of securing the funds necessary for making a start, give the sons of rich families a great advantage over the sons of poor families in making money for themselves.

It is, of course, true that the use of money undermines the foundations of certain causes of disparity of riches that once were potent. Prowess as fighting men, skill as priests, birth as aristocrats, have been the foundation for riches under other forms of economic organizations. Both the history of Europe in the last millenium and the history of savage and barbarian communities in the nineteenth century show that increasing use of money is subversive of economic distinctions established on such foundations. The soldier, the priest, and the aristocrat may be one and all poor men under the new régime; if they are not it is because they have been able to turn their old advantages to a business use. But though the use of money undermines the foundations of these old disparities, it supplies new foundations of the sort described in the preceding paragraph on which diversities are built again.

Whether the new diversities are greater in degree than the old matters little for the present purpose. What does matter is that they are largely the result of the use of money. Since this is the case the use of money must be held responsible, not only mediately but also in large measure ultimately, for the present misdirection of productive energy.

(6) There is another aspect of this question concerning the kind of goods produced to which attention is more often called. Publicists often assert that many of the goods now produced are artistically inferior, or injurious to public health or public morals. "Yellow" newspapers, "dime" novels, vaudeville "shows," alcoholic beverages, adulterated foods and drugs, are familiar examples of goods condemned on such grounds.

This chapter is not concerned with the legitimacy of such criticisms from the point of view of their authors, or with the sociological significance of the fact that these points of view commend themselves to public opinion. The pertinent point of these criticisms is that the kind of goods produced and consumed by society depends on the taste and the ability to discriminate possessed by the people who have money to buy with. If the public prefers melodrama to Shakespeare and patent medicines to the services of physicians, it is futile to blame the "theatrical trust" and the manufacturers of nostrum. If the public cannot tell the difference between pure foods and adulterated foods, it will be supplied with the cheaper article, unless public authority interferes with the ordinary working of the economic organization. Where production is left to the guidance of pecuniary profits dependent on market prices, there men will be found to make whatever goods the community will buy at remunerative prices.

It is also true that under such conditions men will not make goods which the community will not buy at remunerative prices. Yet there are various kinds of goods of great importance for public welfare whose production is not pecuniarily profitable. At the present time the production of such goods is dependent on one of two supports – appropriation by government or endowment by private munificence. Such is the case, for example, with public instruction and scientific research. Both of these methods of providing for such goods leave much to be desired. For it is often difficult to persuade a governing body of the social importance of undertakings that are not pecuniarily profitable, and the man who has money to give is seldom a good judge of what goods the community most needs. Whatever criterion of importance is applied, a careful critic of present production would probably conclude that not enough productive energy is directed to making goods or performing services which are commercially unprofitable.

(7) The use of money affects the determination, not only of the goods produced, but also of the men to whom the direction of the process of production is entrusted.

In business enterprises of the typically modern sort direction is virtually vested in two measurably distinct classes – proximately the class of entrepreneurs, ultimately the class of investors. Other classes, such as manual or clerical laborers, professional men, and landlords, have little part in the direction, except they become entrepreneurs or investors themselves, or act as expert advisers of the latter classes. Their cooperation in business enterprises is essential; but the determination of what enterprises shall be undertaken, and the way in which they shall be conducted, seldom lies in their hands.

Economists have been slow in recognizing the part that investors have in directing production.[3] The reason is partly that the two roles of entrepreneur and investor are often played by the same person even at present; and partly that the role of the entrepreneur as director is much more active and therefore more conspicuous. But in most enterprises the entrepreneur, whether operating on a large or small scale, uses funds belonging to others as well as funds belonging to himself. Whenever an entrepreneur appeals to capitalists to join in his enterprises, to banks to lend him funds, or to the public to buy his stocks and bonds, he gives investors a veto-power over his projects. In deciding to which of the entrepreneurs bidding for the use of their funds they will lend, investors make the ultimate decision as to how society's productive energy shall be used.

The relation of the use of money to this dual direction is partly that of means by which the direction is exercised, part that of the contributory cause of the whole system. The investor's share in directing industry is due to the fact that control over certain sums of money is necessary under the modern system for initiating and carrying on any business enterprise, and that these sums are frequently larger than the entrepreneur can furnish. The entrepreneur's share in direction is due to the fact that business enterprises have become so complex in internal organization and external relationships as to require the constant attention of men highly specialized skill. The work of business management has been clearly differentiated from the work of mechanical operation, and in the larger enterprises is done by a distinct kind of experts. Direction is given to these business experts rather than to technical experts because the existing economic organization makes money profits, not efficient production, the end of activity.[4] Production is more and more carried on in large enterprises where the entrepreneur and the investor clearly divide the direction, because business experience shows that the large enterprise can make greater relative money profits, sometime because it can produce at less money costs, sometimes because it can exert greater pecuniary pressure on competition. Finally, the use of money has provided an indispensable condition for the development of this whole system of elaborately organized production, and in this sense is one of its causes.

The effects of this method of direction on efficiency are in some respects favorable, in others unfavorable. Since anyone who possesses funds available for use in business can to that extent control the use made of productive energy, and since such men lack both business ability and capacity for taking good advice, a considerable amount of productive energy is wasted every year in unremunerative ventures. Probably the division of authority between two classes of directors has some influence in reducing waste, because in many cases the entrepreneurs who form plans have to submit them to a cautious review by investors. Bankers and large capitalists perform important services for society in this way, for they usually bring native sagacity and trained judgment to bear in deciding what plans to support; but the same can hardly be said for the "investing public." And we know from experience that there come now and then periods of speculative excitement when entrepreneur forgets sagacity and investors caution, with the result that the waste of energy in unremunerative ventures becomes enormous. The system is unfavorable to efficiency also in so far as it restricts the number of candidates from among whom the directors of industry are chosen. The would-be entrepreneur of today must have considerable funds of his own, or he must have the support of investors who can provide the funds he lacks before he can secure a trial. These are requirements much easier for the sons of the rich to fulfil than for the sons of the poor. Business ability of a high order is at once so rare and so important in our elaborately organized system of production that this handicapping of the great majority of men must be regarded as a serious defect.

But though the system permits waste by making it easy for incompetent rich men and hard for competent poor men to secure trials as directors of production, it is all the time eliminating the incompetents from among those who are tried, and giving increased opportunities to those who succeed. The incompetents are deprived of the opportunity to waste productive energy by the loss of their money and their credit. The competent men, tho starting perhaps on a small scale, are enabled to extend their control as they succeed. But it must not be forgotten that these men who succeed and achieve important stations among the directors of industry, succeed, not because they utilize productive energy to the best advantage, but because they make money. As has been shown, money can often be made in ways that contribute nothing to or even subtract something from the volume of goods produced; and even when attention is devoted faithfully to secure efficient production the aim is to produce, not the goods that society most needs, but the goods that are most profitable to the producers.

(8) That entrepreneurs often plan and investors sanction business enterprises which result in loss of money to them and waste of productive energy to society is not due solely to lack of business ability on the part of our directors of production. In large part it is an inevitable result of a difficulty presented by the economic organization itself – namely, the difficulty of adjusting production to consumption.

This difficulty of the economic organization is the defect of a merit. The use of money has increased productive efficiency enormously by making possible an elaborate division of labor; but this division of labor places an increasing distance between the men who make goods and the men who use them. Though business men are compelled in the interest of their own profits to devote much time and sagacity to the unremitting study of consumers needs and whims, they find it impossible to forecast with accuracy the changes that are all the time occurring. Thus an important element of chance is introduced into most lines of business. Even when goods are made only "to order," the producers may make disastrous mistakes by investing large sums in enterprises which can't be kept after they are organized. Every time a factory is built in anticipation of a demand that does not materialize; every time goods are made that can't be sold except at heavy loss; every time that the directors of industry underestimate the quantity of certain goods that the community requires, there occurs a misdirection of productive energy. Such mistakes are not accidents, they are certainties produced by the economic organization.

The ill effect of this element of chance in all business calculations is more far-reaching than at first appears. The very fact that a businessman realizes the impossibility of foreseeing all contingencies that may affect his plans, betrays him into relying more than he need on chance and less than he might on judgment. A man's pessimistic or optimistic temperament, his temporary mood of caution or confidence, often weigh more on his decisions upon business venture than cool calculations. Thus the element of chance imported into most business enterprises by the wide separations of producers and consumers makes the direction of productive energy depend largely on states of feeling. This fact contributes toward making business crises and periods of depression. A man's state of feeling is powerfully influenced by elation or depression among his associates. Hence, entrepreneurs and investors usually exhibit at any time a certain consensus of feeling that leads them to a certain uniformity of action. This uniformity of action is increased by the fact that action based on a state of feeling frequently produces business conditions that give rational ground for similar actions by others. Confidence is not only infectious; it also produces a condition of business that justifies confidence – up to a certain point.[5] The result is that ill-considered enterprises are buoyed up for a time and similar new ventures are encouraged until the maladjustment production and consumption becomes serious. Finally, the least successful among the many tottering enterprises fail and their failures frighten many men. The state of confidence quickly changes into a state of distrust which, like its predecessors, is directly infectious and also incites actions that produce a business situation which justifies the feeling. These circumstances draw into the circle of disaster many enterprises which at more favorable times would be profitable to their directors and serviceable to society. The whole complex business organization is thrown into disorder, – a disorder which inevitably involves the process of production, because production is carried on by business men for business ends.

(9) Finally the basing of economic relations upon money prices exposes the whole process of producing the goods required to meet social needs to disturbances arising from the technical exigencies of monetary and banking systems.

Though the existing economic organization constrains the individual to think about his economic interests habitually in terms of dollars and cents, these monetary units are to him but symbols representing in a convenient shorthand the variety of goods that he can get if he gets the money. But as such symbols monetary units have a serious defect; the amount of goods which they represent is subject to continual changes. For example, a man who agrees to work for a certain salary may find that his anticipations of the comfort in which he can support his family with this amount of money are sensibly falsified by a change in the money cost of living.

From the standpoint of the individual most changes in prices are arbitrary conditions to which he must accommodate himself as best as he can; they are beyond his control and ordinarily cut little figure in his calculations, except they affect commodities in which he deals as a business man. But from the standpoint of society price changes are an inevitable outcome of the economic organization itself. For this organization makes money prices the battle-ground of conflicting economic interests, and control over prices the chief weapon of economic aggression and defense. Every circumstance that affects the relative business position of buyers and sellers of a good results in a change of price.

And every change in the price of an important good is likely to lead to repercussions which result in changing the prices of other goods. A man injured by advances in the prices of goods he buys seek redress by demanding a higher price for what he sells, and if he succeeds he puts pecuniary pressure on another set of men to alter the prices of the goods in which they are interested.

Besides these price changes growing out of the shifting relations of demand and supply of goods there are other changes growing out of shifting conditions in monetary and banking systems. Where the monetary system is based upon the use of gold and silver coin the scale of prices is exposed to changes arising from such extraneous circumstances as the discovery of new or exhaustion of old mines, improvements in methods of smelting, political disorders in the producing districts, etc. Again, prices are frequently thrown into confusion by attempts on the part of governments to gain a respite from financial embarrassments by debasing the coin or issuing irredeemable paper money in its stead. Even in the nineteenth century not one among the most important nations of Christendom escaped a misadventure of this sort. Once more, in those nations which make large use of bank notes or checks as a circulating medium the course of prices is independent on the exigencies of the banking situation. That business men are aware of this fact is shown by the attention with which they follow in all parts of the United States the weekly reports of the Associated Banks of New York, and even the cabled reporta of the great banks of Europe.

Changes in the prices of particular articles and changes on the general level of prices, then, are all the time occurring. Since the economic welfare of the individual depends on the prices he receives and the prices he pays, these changes introduce an element of uncertainty into his economic life, and since the production of wealth is carried on to secure money profits and profits depend on prices, these changes affect the rate of production – sometimes stimulating it, sometime depressing it. In other words, men have not attained complete control over the social mechanism of prices; in part the mechanism controls its makers and warps their activities to suit its exigencies.

NOTES

1. Cf. Thorstein Veblen, *Theory of Business Enterprise*, New York, 1904, pp. 35–41.
2. Natural Value, tr. by C. A. Malloch, London 1893, p. 38.
3. Perhaps the clearest statement on the subject is that of Professor F. W. Taussig in *Wages and Capital*.

4. Cf. Thorstein Veblen, *Industrial and Pecuniary Employment* (paper read at the 13th Annual Meeting of the American Economic Association), 1900.

5. Cf. E. D. Jones, *Economic Crises*, New York, 1900, Ch. IX.

REVIEW ESSAYS

Multiple Review of van Creveld's THE RISE AND DECLINE OF THE STATE

THE SHIP OF STATE IN STORMY WATERS

Y. S. Brenner

London and New York: Cambridge University Press, 1999, pp. viii, 439. Index. ISBN-0-521-65190-5. $54.95

Professor van Creveld feels that the state is in decline. He says that regardless of whether states fall apart or combine many of their functions are being taken over by a variety of organizations which, whatever their precise nature, are *not* states (pp. vii, viii). He defines the state as an *abstract* entity. An entity which is not identical with either the rulers or the ruled, nor even with an assembly of all the citizens acting in common, but includes them and claims to stand over both (p. 1). "It possesses a legal persona of its own, which means that it has rights and duties and may engage in various activities *as if* it were a real flesh-and-blood living individual. The points where the state differs from other corporations are, first, the fact that it authorizes them all, but is itself authorized (recognized) solely by others of its kind; secondly, that certain functions (known collectively as the attributes of sovereignty) are reserved for it alone; and, thirdly, that it excercises those functions over a certain territory inside which its jurisdiction is both exclusive and all-embracing" (p. 1). The most important characteristics of the state are: being sovereign, being territorial, and

Research in the History of Economic Thought and Methodology, Volume 19A, pages 105–142.
ISBN: 0-7623-0703-X

being an abstract organization with an independent persona which "is recognized by law and capable of behaving *as if* it were a person in making contracts, owning property, defending itself, and the like" (p. 416). Defined in this way, which is more or less the same definition as adopted by Benn (Benn, 1967) and Sabine (Sabine, 1953), and without claiming that his definition is exclusive, Professor van Creveld finds that during most of history there was governance (he uses the term government) but no states, and taking this as his point of departure he makes a brave attempt to look into the future of the state by examining its past. The term governance would be more appropriate in this context because *government* usually denotes a kind of polity ruling and directing the affairs of a state, whereas *governance* means the action or manner of control – the method of management or system of regulations, of an association that need not neccessarily be a state.[1] Professor van Creveld believes that the man who really "invented" the state was Thomas Hobbes; and that from his time up to the present, one of the most important functions – as of all previous forms of political organization – had been to wage war against all others of its kind. He says, that had it not been for the need to wage war, almost certainly the centralization of power in the hands of the great monarchs would have been much harder to bring about, and if it had not been for the need to wage war, then the development of bureaucracy, taxation, even welfare services such as education, health, etc. would probably have been much slower; and that in one way or another all of them were originally bound up with the desire to make people more willing to fight on behalf of their respective states. He does not share the views of those who believe that the sovereign state is the root cause of war, but rather that the real reason why war exists is because men have always liked war, and women, warriors, and that states can develop a strong appeal to the emotions only so long as they prepare for, and wage, war. If, for any reason, they should cease to do so, then there will be no point in people remaining loyal to them – their raison d'être will have been lost. Hence, since nuclear weapons have since 1945 diminished the ability of states to go to war, the state is in decline (pp. 336–337).

The book is divided into six chapters and conclusions. The first chapter deals with the period when there were no states. The second (from approximately 1300 to 1648) discusses the era before the state became an abstract organization with its own persona separate from that of the ruler. The third chapter (from 1648 to the French Revolution) examines how the state was gradually becoming an instrument to impose law and order and to safeguard life and property and how in the process it built up a bureaucracy, created an infrastructure, monopolized violence, and developed a political theory. The fourth chapter (the years separating the French Revolution from the end of

World war II) explains how the forces of nationalism transformed states into secular gods fighting each other for supremacy. Chapter five takes the reader backward and forward in time portraying the spreading of the state from its original home in Western Europe to other parts of the globe. Chapter six deals with the forces that since 1975 were undermining states, and raises the problem of the dichotomy between government and governance.

Governance in the simplest communities began and ended within the extended family, lineage, or clan (pp. 2–7) Chiefdoms introduced hierarchy, permanent authority, domination of one group over another and were the first political entities to impose compulsory unilateral payments (rent, tribute or taxation) which concentrated wealth in the hands of the ruling few. (pp. 17–19). They were overwhelmingly rural. In cities, both in the Greek *polis* and in the Roman Republic, citizens did not constitute a single body but were divided into "demes", "curies", "centuries", and "tribes" which, in Rome at any rate, cast their votes en bloc.

In Greece and Rome governance (*arche, imperium*) was a form of authority exercised by some persons over others who, unlike family members and slaves, were equals before the law. There was a sharp distinction between the private and the public spheres. In the house control was exercised by the *pater familias*, outside the house there was a political authority, or governance. But the city-state's organs of government did not correspond to our customary separation between executive, legislative, and judiciary (pp. 23–25). The normal method for meeting the expenditure of government in peacetime was to rely on market duties and income from the justice system such as fines and confiscated property. Most important were the *liturgies*, contributions made by the richer citizens for specific purposes as a civic duty but not really voluntarily.

The early empires were mighty organizations. Some, like the ancient Egyptian and Chinese empires lasted for centuries. At their head stood single rulers most of whom claimed to owe their position to a divine connection. They were absolute rulers who combined legislative, executive, and juridical functions and being something like divine claimed universal sovereignty and regarded other political communities as troublesome barbarians. Apart from religion the two pillars supporting the imperial rule were the army and the bureaucracy. But the two were sharply differentiated from each other to avoid collusion between them which could pose a risk to the emperors. Taxes collected from the public belonged to or were at the disposal of the emperor in person and so were the army and bureaucracy. The members of the administration and of the army owed their allegiance to the emperor in person and the absence of a clear distinction between the private and the public spheres left almost the established religion or church alone a more or less safe institution against arbitrary interference from the emperor's absolute power (pp. 36–46). Arbitrary, in the sense of uncontrolled, based on personal opinion, or used without considering the wishes of others.

Feudalism gave rise to the placing of increasingly greater emphasis on the collective rights of the aristocracy and the religious establishment and subsequently to the idea that government was an integral part of the divine order in which each person and class has an appointed place protected from arbitrary interference. Privileges became attached to their holders and acquired certain "constitutional" guarantees.[2] Once the warrior-governors ceased to be appointed by the emperor and succeeded in making their positions hereditary, the process reached its logical conclusion and however alive it still was in name the empire came to a de facto end (pp. 49–51).

The demise of the empire and the birth of states was attended by three centuries and a half of struggle for preeminence between monarchs and church, nobility and towns. During most of the Middle-Ages the pope in Rome was beyond the reach of the emperor and any other secular ruler. The church inherited the language of the Western Roman Empire and its legal and political traditions. As the church's assets were scattered all over Europe it developed a sophisticated financial, judiciary, and administrative apparatus capable of overcoming distance and time, which by 1300 was far in advance of anything of the sort available to secular rulers. As for centuries it exercised a virtual monopoly on literacy the church's services became indispensable to any secular ruler with extensive domains. By 1302 it was at the climax of its power and Pope Boniface VIII, proclaimed that the secular power should be exercised at the command and sufferance of the priest. The centrifugal tendencies inherent in feudalism were at work for the best part of a millennium but the final collapse of the Roman Empire was delayed until the rise of the great monarchies (pp. 57–61).

The struggle between ecclesiastical and secular governance continued well into the fifteenth century when the church's secular authority was challenged by the new humanist scholarship and the most important monarchies were already well established and the church's former justice system was curtailed and financial independence and property often subjected to royal taxation (p. 66). But the most powerful blow came with the Reformation. Weakened by Wycliffite, Hussite and earlier rebellions, blemished by the Babylonian Captivity and the Great Schism, its prestige marred by corruption and worldliness, and shaken by the critical scholarship of Erasmus and others, Rome's hegemony over the minds and morals of men did not survive the onslaught of the rising nation states, capitalist economies, and of the growing middle classes which felt restrained by it. Once the princes' fears concerning the political effects had been assuaged, Protestantism spread rapidly. Church's property was seized and priests were either turned into royal servants whose duty happened to be looking after peoples' souls. As secular rulers tightened their grip on the church, change also overtook the personnel of government. By the beginning in the 15th century, encouraged by the spread of secular humanism, laymen were increasingly able to get as good an education as ecclesiastics and rulers became gradually less dependent on "clerks" to staff their administration (pp. 66–70).

Sometime before 1517, Machiavelli expounded in the *Discourses* the first modern theory of politics and suggested that one of the secrets of political stability consisted of the upper classes using religion in order to keep the masses in their place. And indeed, from about 1600, both Catholic and Protestant rulers were beginning to treat their churches as if they were mere departments of state. While most educated people probably continued to believe both in the divine right of kings and in the latter's right and duty to look after their subjects' spiritual welfare, change was in the air (pp. 71–72). In the words of R. H. Tawney (Tawney, 1922), (who is not mentioned by Professor van Creveld) this was an era in which an increasing number of people turned "from spiritual beings who, in order to survive, devote reasonable attention to economic interests, into economic animals who also find it prudent to take some precautions to assure their spiritual well-being". Perhaps the most radical view with regard to the position of the ruler was, according to Professor van Creveld, in Thomas Hobbes' *Leviathan* (published in 1651) in which he declared God, if He existed, irrelevant to politics; and recommends that subjects be made to practice the religion prescribed by the sovereign as best adapted to the maintenance of public order (p. 73).

Although the kings were winning the struggle against the church, the Emperor of the Holy Roman Empire retained his titular position as the head of the feudal hierarchy. This position was not without some practical significance because he was the only one who could create kings. Moreover, during the last quarter of the fifteenth century the idea of empire as a universal organization was alive in popular consciousness as the force to protect the Christian world from the advancing Turks (pp. 77–80).

On his installation as Emperor in 1519, Charles V, had to swear a coronation oath which was formulated in a manner that for the first time allowed the princes a voice in the administration of Imperial affairs. But the watershed which marked the monarch's triumph over both Empire and the Church, and led to the partition of the Imperial territory, was the peace of Westphalia in 1648 (pp. 81–86).

As the fight against the universality of church and Empire was being won by the monarchs, they were also making headway in their efforts to restrain feudal particularism. Slowly the crown's competitors, first the cities and then the nobility, were turned into the monarchs' associates (pp. 87–117). Yet, neither nobility nor cities were entirely defeated. The former retained privileges and a near-monopoly on the upper ranks of government, and the latter attained the support which allowed them to flourish economically. In western Europe Capitalism and monarchy marched together side by side. Capitalism provided the monarchy with financial muscle and monarchy provided capitalist enterprise with military protection within and later also outside its borders.

Between 1648 and 1789 Western Europe weaned itself from the imperial traditions and the state emerged as an instrument of governance while the person of the ruler became separated from his "state". Indirect rule by feudal lords changed to direct government exercised by salaried officials on the king's behalf. Professor van Creveld traces this development in four spheres, namely in the rise of the bureaucracy and its emancipation from royal control; in the strengthening of its hold over society by the reorganization of the collection of information and taxation; in the way bureaucracy and taxation together made it possible for the state to create armed forces for external and internal use and to monopolize the means of violence; and by the evolution of a political theory which accompanied and justified all these developments (pp. 127–128).

By the early 17th century two kinds of personnel could already be distinguished. One which derived its authority and income from the possession of land and from the personal feudal nexus, and another which was appointed by the kings whom they served with or without remuneration, and who could be transferred, promoted, and dismissed at the monarch's will. The move from the feudal system of administration, through what may be termed state entrepreneurs, to appointed salaried officials led to a shift from a geographical to a functional division of labor. In the process did not only the number of officials increase spectacularly but also the amount of paperwork. The invention of the printing press made royal decrees and ordinances available to anybody who wished to consult them (pp. 128–136).

By the middle of the 18th century, officials, who for centuries had been the king's men, were beginning to think of themselves as servants of an impersonal state. The king's officials became "flunkies" and the most important of the state officials became ministers. Though the details varied in the different countries, the century and a half after 1648 may well be characterized as the era of the growth in the power of state bureaucracy which by the beginning of the 19th century had become the master over civil society. What Professor van Creveld had in mind mentioning civil society is not quite clear here. In the period he is dealing with in this part of his book it was used by liberals for the purpose of attacking absolutism. In modern social and political philosophy it is usually related to human activities and to a set of institutions outside state or government, such as families, voluntary associations and social movements. Recently, the expanded role of the state in economic and social life seems to deprive the concept of both its intellectual home and critical force.[3] However, in the process of becoming the master over whatever is implied by the term, an administrative infrastructure emerged: maps were drawn and statistical information was gathered on population, property, production and incomes. With this the tasks of government and the number of bureaucrats expanded and the rulers' private resources dwindled into insignificance in comparison with the state's, and rulers ceased to be personally accountable for the government's debts. Gradually taxation also changed from indirect to more remunerative direct taxes. The granting of state monopolies, the abolition of traditional privileges and tax exemptions, spiked further the states' revenues. Naturally, none of this could have happened without the administrative reforms, nor without the wider spreading of education, to replace class, property, and connections as a means for personal advancement, nor without the state taking control of the means of violence to prevent foreign invasions and suppress international opposition (pp. 137–155).

The transition from feudal hosts to mercenary forces and thence to regular, state-owned armies and navies, made its appearance after 1648. The Thirty Years War marks the beginning of the new system of state monopoly of the means of violence (p.159). After the Treaty of Westphalia war was no longer waged for personal reasons and many of the mercenary forces that had fought the war were absorbed into the standing armies or *militia perpetua,* while companies trading abroad came increasingly to be regarded as an extension of their governments' power (pp. 156–161).

By 1700 the uniformed guardians of the state divided into forces for external conflict and for internal order – into army and police. War was coming to be the continuation of policy by other means, and the attempt of lesser groups to use violence to further their ends were termed rebellion, banditry and later terrorism (pp. 165–170).

Together with these practical changes came a transformation in the way people thought. The first to define the state as an "artificial man" separate from the person of the ruler was Thomas Hobbes. Hobbes was writing in a period of civil wars and with the object to restore order. In his view there were two kinds of "bodies" natural bodies, such as man himself, and artificial bodies, who could be private or public. The private were formed by individuals on their own initiative and the public bodies were created by the state. In this system the state was the more important. It authorized the former while it itself was authorized by none. With this Hobbes created the state (the commonwealth) as an abstract entity separate from both the sovereign and the ruled. The latter transferred the right to govern to the ruler by means of a contract among themselves. The ruler could be both a single person or an assembly whichever was more convenient. But because "covenants without swords are but word", the ruler had the power to impose the law. Hobbes saw man as basically evil and man's reason motivated by fear so that he spends his entire life seeking power vis-à-vis his fellow men in a struggle ending only in death. "During the time men live without a common power to keep them all in awe, they are in that condition which is called war; and such a war as is of every man against every man . . ." (Hobbes, 1651). It was to restrain this creature that Hobbes set up the sovereign and made him the most absolute in history (pp. 179–180).

John Locke's (1632–1704) work in the field of politics must be seen as a reply to Hobbes. Hobbes' human essential quality of rationality led man to war of all against all and hence to the need to create a sovereign to restrain this war. Locke saw man's rationality in terms of an enlightened self interest which leads people to live even in the state of nature most of the time in relative harmony. His social contract was therefore not designed to safeguard people against their ill nature but to improve their life – safeguarding their natural endowments of right to life, liberty and property. The thing to be avoided was *absolute* government. Government was to be based on consent repeatedly confirmed by elections. He did not explicitly explain how, but suggested to divide government between a legislative, an executive and a federative authority charged with the conduct of war and foreign policy (pp. 180–181). Montesquieu tried to discover ways to protect civil society against the arbitrary power of the sovereign. According to him tyranny was not avoided by the law of nature but by laws made by man for himself. Laws which were written down in accordance with the kind of community man had in mind. The standard by which communities had to be judged was the degree of liberty. Good or bad was only what the state enacted (pp. 181–182).

At this point Professor van Creveld ignores the entire awakening of modernism which began with Galileo and Descartes and takes the reader directly to David Hume (1711–1776) saying that reason may well not be more than the servant of passions, and that it was not only manifestly untrue that such a thing as objective reason can be shared by all people, but even if it could any connection between it and the intentions of nature would be completely undemonstrable.

Between Hobbes and Locke, van Creveld postulates the theoretical structure of the modern state, in which an abstract entity separate from both ruler and ruled but including both, divorced from both God and nature, and no longer bound to observe custom except insofar as it had been ratified by itself, and capable of doing anything, was substantially complete. Between about 1500 and 1648 the state was not conceived as an end but only as a means. Its overriding purpose was to guarantee life and property by imposing law and order while things like gaining the consent of the citizens and securing their rights, were still considered secondary. It recruited its administrators and officers from the upper classes; took taxes from the middle ones, and from the lower strata it obtained both taxes and cannon fodder. Yet the real great transformation was still only in the offing (pp. 182–189).

Professor van Creveld's thinks that the works of Jean-Jacques Rousseau (1712–78) did probably more than anyone else's to start the Great Transformation. For Rousseau individuals' mental and moral formation took place in the community. In the *Social Contract*, (1762) he suggested that this community had a corporate persona represented by the general will. The submission to, and participation in, this general will was patriotism – Patriotism was virtue. But Rousseau was no nationalist, though this was the period when the reaction to the universalistic ideas of the Enlightenment were beginning to come under attack from the early Romantics who were denouncing the Enlightenment's emphasis on the rational and uniform (pp. 191–192).

The early Romantics claim was that, nurtured by soil and climate, every nation has its own culture and character, which manifested itself in dress, habits and language. Yet eighteenth-century intellectuals were not nationalists in the modern, political sense. They regarded national cultures as a garden of separate flower beds with each with its own value. The Napoleonic victories in Germany (in 1807–1808) brought about a drastic change. In Germany, Fichte raised the anti-French sentiment to an almost religious fervor. His works indicate the move from German cosmopolitism and inclination toward pacifism in the direction of the militant chauvinism that was to stay into the twentieth century (pp. 193–195). Hegel's *Weltgeist* attributed reason to the separate communities or states in which they lived and endowed states with historical importance, and sovereignty with an ethical content. True freedom for an individual was only possible within the state. Take out the state and man becomes a puny biological creature whose life is divorced from the world-spirit and in this sense devoid of ethical significance. With him the state was the community's highest, indeed sole, representative.

From the 1830s the way to play out its historical destiny was for the state to pit itself against other states by means of war. In German minds this became the principle tool whereby the world-historical spirit unfolded itself, and without which everything tended to sink into selfishness and mediocracy. In France and

Italy men like Guizot and Mazzini believed in combining national greatness, independence and power with personal freedom for the individual. Professor van Creveld says that these were the ideas that fueled the interstate rivalry which was the prominent feature of the period from 1848 to 1945 (pp. 195–197).

The first state to mobilize the masses was Revolutionary France. Between 1789 and 1815 France set up a centralized bureaucracy, established general military service, produced a comprehensive legal code with authority over all Frenchmen regardless of status, creed or province of residence, and introduced state directed secondary and tertiary education. To popularize the idea of the nation as a supreme value the Republic initiated huge popular festivals to celebrate itself, introduced a new national flag and an official anthem. Soon other countries followed the French example. In this way was the nationalist cause married to the state until state worship reached a point where the original distinction between it and civil society disappeared, and in the end made *war* the crucible of the nation as well as of the state in which it organized itself (p. 198). Between 1815 and 1860 only in countries where nation and country did not coincide did rulers still have cause to fear popular nationalism and did not encourage it. This was the case in pre–1866 Germany, and in the Austro-Hungarian Empire.

With industrialization and the growth of towns people were deprived of their traditional rural social safeguards, and feeling alienated they found a surrogate sense of "belonging" in their national identity. Before long all manner of activities evolved as part of this national personality – even sport. Voltaire still thought of patriotism as a scoundrel's last refuge, but after 1789 only socialists doubted that the highest human virtue was patriotism. This process of deification of the state and the attending patriotism was consciously promoted by education on the one hand and by the use of police and prisons to suppress dissention on the other (pp. 198–201).

The success of the modern state hinged on its willingness and ability to protect the property of its supporters. Marx recognized this and so did the anarchists, but until close to the end of the 19th century their objections were mostly voices in the wilderness. Not that people were unaware of the "social question", but those in power looked upon it more as a threat to the foundations of the established order and to the labor discipline that modern capitalism required, than as a just claim for equity and everyone's right to a share in welfare (pp. 206–207).

A most important role in the conditioning of society to regard nationalism and discipline as supreme values was played by education (p. 211). States were introducing schools and state officials determined the curricula. Education established a firm grip on the minds of the young. By and large, the early 19th century had been the heyday of laissez-faire. But already in the 1830s the direction began to change. Professor van Creveld tells the reader little if anything about the role of Trade Unions and other labor movements in bringing about this change. The reasons he gives for the social legislation are to acquire the allegiance of those old enough to perceive that their real interests consisted not of circuses but of bread, and that (p. 211) some reforms were out of concern for the welfare of society and others just for fear of revolutionary disturbances. (pp. 211, 217–218). He ignores the rising power of the social democratic labor movements, the piecemeal achievements of Trade Unions and the impact of the early days after the Russian Revolution on workers and capitalists, and just mentions the kind of Bismarckian ruling on working conditions and health care, in the belief that the "*political* redefinition" of the character of the state in the western world only began during the Great Depression with the New Deal when a solution had to be found to the massive unemployment, when the state had conquered the currency and thus brushed away the financial constraints on waging war and the pecuniary limitations on the its ability to sustain a police force, education system, and social services (pp. 217–224).

With this, Professor van Creveld says, the bureaucracy, "born in sin, the bastard offspring of declining autocracy" was running amok, and the state became a giant wielded by pygmies. As individuals, he explains, bureaucrats may be harmless people, but collectively they have created a monster whose power outstrips that of the mightiest empires of old. They do not pay the expenses of government out of their own pockets as used to be the case before, and they work according to fixed regulations and procedures without anger or passion, though they favor their own interests above all. But most important they possess a collective personality which makes them immortal. Together with the bureaucracy state dominance grew and grew and stage by stage the state separated itself, and raised itself above, civil society.[4] It commissioned maps and used them to make political statements about itself; it built up an infrastructure of statistical information; it increased taxes, and concentrated them in its own hands. To complete its dominance, it set up police and security forces, prisons, armed forces, and specialized organs responsible for looking after education and welfare which reflected the mechanism it served, and spread its wings over money. The main difference between "free" and totalitarian states consisted in the former chose their rulers by democratic elections and did not have to be ruthless to coerce people to accept them (pp. 257–259). Here Professor van Creveld forgets that in this sense Nazi Germany would have to be included among the "free" states at least initially.

In summary, first conceived as a mere instrument for imposing law and order almost exactly midway in its development between (1648 and 1945), the state came across the forces of nationalism which, until then, had developed almost independently of it and sometimes against it. As conceived by Rousseau and Herder, "nationalism" had been harmless preference for one's country, language customs, dress and festivals; but once it had been adopted by the state, it became aggressive and bellicose. Digesting stolen spiritual goods, the state turned itself from a means into an end and from an end into a god. In peace it extended the right to the same godly sovereignty to other countries, but from its own subjects it demanded absolute loyalty unto death and inflicted savage punishment on them who dared to disobey. Protected and often abetted by the state, modern science and technology flourished. Science and technology augmented standards of living for which the state exacted protection money in the form of high taxation, and enhanced its military capabilities for combating other states and to reinforce its grip on every inch of territory and the life of every individual of its citizens. And so, according to Professor van Creveld, aided by the existence of the popular press focused on its own interests the state transformed itself into god on earth (pp. 259–261).

Surprisingly Professor van Creveld does not mention the forces working in the opposite direction. He ignores the fact that the last decade of the 19th century saw the birth of the many revolutionary international forward-looking movements which by the 1920s were triumphing everywhere in Europe. What all these movements had in common, from Surréalisme, Cubism to Vorticism, and from the international scientific cooperation of people like Einstein to most of the original thinking now called Modernism, can hardly be dismissed as the state-directed bogey which Professor van Creveld apparently believes to have dominated this period. Was there no International Brigade in Spain trying to stop Franco? Was there really no international humanist literature before, and even during, the high-tide of Fascism in Italy, Nazism in Germany, and Stalin's subverted Socialism in Soviet Russia and in the rest of the nation states? There was! Even in Germany, prior to the Nazi's coming to power there was a vast public which was far off from cherishing the nation-state. The fact is well reflected in the literature produced by men and women like Stefan Heym, Heinrich Mann, Bertolt Brecht, Fritz von Unruh, Anna Seghers, Johannes R. Becher and all the many other progressive German authors, and like Thomas Mann, Hermann Hesse, Feuchtwanger, Zweig, Werfel, Holthusen and later Borchert, in the moderate center, who were disgusted with both the nationalistic chauvinism and with the self-satisfied bourgeois mentality, and who were not prepared to abandon the rest of the liberal bourgeoisie's heritage.

In Chapter 5, Professor van Creveld surveys the spreading of the state beyond the locations of its origins. He locates this in the period between 1696 and 1975. He begins with Russia. While the West was undergoing the tremendous upheavals of the industrial revolution, and experienced repeated outbreaks of revolutionary violence in 1830 and 1848–1849, social and economic change in the gigantic Russian empire proceeded at a glacial pace, but the bureaucracy expanded. The Crimean War (1854–1856) disclosed Russia's backwardness and showed that it needed reform. With the separation of the judiciary from the executive in 1861–1864 arbitrary government was brought to an end in at least one crucial respect: serfs were given an independent legal persona and the right to own property and against a quit-rent payment to move to town. The freedom was limited but from 1870 on sufficient social mobility was created to help industry to take off. Funded by the treasury, a gigantic railway-building program was launched, and from the 1890s came a spectacular expansion of heavy industry, much of it also state financed to serve its needs. A small civil society finally emerged around 1900. Most of the intelligentsia consisted of educated property owners (lawyers, teachers, low-level officials, students) and some aristocrats. Among these were liberals who admired the West, Anarchists who saw in the state the root of all evil, and Slavophiles who rejected modernism. Too small to have real influence the intelligentsia turned to seek an alliance with "the people".

The war with Japan (1904–1905) demonstrated again the weakness of the tsar's rule. The abortive revolution that followed, and the last minute attempts to broaden the regime's support by democratization, faltered. The masses became increasingly radicalized and by the time of Word War I (1914–1918), they were ripe for revolution. The Communist polity established after 1917 did away with the last vestiges of patrimonialism, and henceforth, for ideological reasons, everybody was turned into a servant of the state (pp. 264–276).

In Poland King Stanislaw II Poniatowski (1764–95) started enacting reforms. Inspired by the ideas of Montesquieu, Rousseau, and Washington, he set up the first modern Polish cabinet which in turn led to reform in the tax system and to the creation of the nucleus of a modern army. The reforms did not save Poland from her second and third partitions (1793 and 1795). An independent Polish state was only resurrected in 1918 when Germany and Austria were defeated, and Russia was undergoing a revolution followed by civil war (1918–1919) (pp. 279–280).

Compared to earlier empires, the British in America arose at a time when the separation between private proprietorship and political rule was becoming firmly established in England. Governors were not "owning" the colonies but taking them under a political régime securing the inhabitants' loyalty to the crown. Governors did not have extensive bureaucracies at their disposal and so, deprived of a proper administrative machine, they sought to avail themselves of their subjects' aid by means of councils recruited from Americans who were mainly prominent landowners and merchants. The rebellion against Britain was therefore conducted by elected representatives on behalf of a nascent abstract state. The Constitution adopted in 1788 was primarily based on the ideas of Locke and Montesquieu, that is on government by consent and designed to protect the individual, his liberty and right to property. The franchise was given to all white, male, tax-paying citizens. The Constitution, and the amendments of 1791, left open the measure of states rights and of the federal government's. Improved transportation permitted the states to mobilize their resources almost as efficiently as small states and helped the development of the USA into a centralized and growing economic power. The question of slavery and tariffs continued to prove a bone of contention between North and South until it was resolved by the Civil War.

Aware of its growing power the USA also began to develop a robust nationalism between the 1880s and 1890s. The exponents of American imperialism did not consider themselves reactionaries. They were self-styled "Progressives" who were usually paying lip service to democratic principles. In practice they were not very different from the European nationalist. The absence of a serious external threat may explain why it took long before a centralized army and a centralized bureaucracy developed in the U.S.A. But eventually the New Deal ushered in a limited change in this respect. In effect it was only World War II, and the subsequent Cold War, which finally introduced Big Government (pp. 281–289).

The Canadian state was shaped by the different pulls of the British and the American model. It resulted in the Dominion model, the model which was equally successful in New Zealand and in South Africa (pp. 295–296).

The fate of Latin America was different. Whereas much of the Anglo-Saxon expansion during the period 1600–1850 took place against the background of almost empty continents, this was not the case when the Spanish and Portuguese colonized Central and South America. There, for example, the conqueror of Mexico Hernando Cortés alone received an estate or *encomienda* with 23,000, Amerindian serfs who owed him tribute and over whom he acted as landlord, governor, supreme justice, and chief of police all rolled into one. Discovered by Europeans at the time when the separation between ruler and state was only beginning, the conquered lands were seen as the property of the king. The highest authority on the spot was that of the royal governors. As in contemporary Europe the bureaucratic pyramid that emerged was venal as officials, having bought their posts, tried to compensate themselves for their investment and make a profit. The towns enjoyed an element of self-government in the form of municipal councils.

Latin America was beset by deep racial divisions. Though the gradations were often absurd, the prejudices behind them were very real. The white arrivals from Spain and Portugal monopolized all important offices. The native-born whites, who were lower level officials, came second, and below them were the poor landless whites, and finally blacks and Amerindians who were almost entirely excluded from offices. Efforts by the end of the 18th century, to reform the system in order to strengthen royal control and reduce corruption by giving more say to the bulk of the population failed (pp. 298–304).

Chile somehow succeeded in maintaining a political tradition unbroken by violence. Between 1830 and 1870 government tended to be in the hands of the conservative landowners and there was no tradition of slavery, serfdom, and proprietary government. When the transition to liberal rule came it was achieved by constitutional means; except for the privileged position of the armed forces and the fact that the franchise always remained rather narrow, the resulting system of government was in many ways like that of the United States (p. 305).

In the rest of South America, later, army officers often sought to impose order. Sometimes they helped themselves to power at the head of juntas made up of fellow officers or former bandit leaders. To the extent that the endless civil wars permitted it, economic development began in the first half of the 19th century. The old imperial power was breaking down, and even before independence the influence of European ideas encouraged liberals in favor of free trade. But political instability prevented the accumulation of capital. When in the second half of the 19th century, the influx of capital, first from Britain and then from the U.S.A, started, foreigners and conservatives found a common interest in political tranquility and cheap semi-servile labor. And so, it can be said that by Professor van Creveld's definition of the state, by the last quarter of the 19th century many Latin American states were states mainly in name alone (pp. 306–307).

Another feature of the last quarter of the 19th century was the beginning of large-scale immigration into what hitherto had been a remarkably under-populated continent. Although immigration brought some benefit to agriculture, most newcomers settled in cities and gave rise to something like modern political parties. In the most important countries this, and a measure of industrialization (that began in 1920s) brought old traditions to an end. In Argentina, after the coup of 1930, armies began to play an increasingly dominant role. Seeing the weakness of the civilian institutions they believed themselves to be the only factor capable of organizing things above the narrow interests of factions and class, and they often developed fascist sympathies. Disciplined or not the long tradition of civil wars and coups meant that a key characteristic of the modern state – a clear separation between the forces responsible for waging external war and those charged with maintaining internal order – did not emerge and with falling wages the cliques found themselves confronted by left-wing labor organizations (pp. 309–311).

In the cities the populations in latin America in the mid-twentieth century were too poor and inarticulate to pose a political threat and sufficiently powerful to create constitutional modern states. In marked contrast to the situation in the U.S.A and the Dominions with a few exceptions none of the South and Central American regions have been able to put their people under the rule of law or establish civilian control over army and police (pp. 311–314).

The last societies to adopt the state as their dominant political entity were those of Asia and Africa. Asia contained some of the most ancient, most hierarchical, and most powerful empires of all times with a bewildering variety of political systems, but there too none seems to have developed the concept of a state as an abstract entity which is not identical with either the rulers or the ruled etc.

As distinct from other parts of the globe, European power in Asia and Africa began to spread from overwhelmingly commercial interests and not from settlers. The "factories" planted on the coasts of Africa and Asia were run for centuries by the various colonial companies who appointed their officials as governors. Between 1600 and 1715 the East and West India companies of Holland, Britain, and France often engaged each other in hostilities while their respective governments remained in peace. Even after these tussles ceased, the companies continued to maintain their own bureaucracies and armies. The shift from commercial ownership to rule by government was a prolonged process and territorial expansion took even longer. However, by 1914 European technical superiority had led to the partition of the entire world between a very small number of competing states (pp. 315–317).

Only Japan, Thailand and ignoring the relatively short period of Italian occupation on the eve of World War II, Ethiopia, escaped colonialization. Colonial rule can be divided between the extremes of Belgian direct rule in the Congo, and British rule which placed greater reliance on African Chiefs. Belgian businessmen recruited workers on plantations and in mines and forced them to stay there. In the British African colonies the chiefs had most important functions of government and in time were absorbed as paid officials into the British system of government. But regardless of the way they were administered colonial governments weakened the traditional native institutions, introduced cash taxes, and created new semi-educated elites. The primary *ideological* justification or excuse behind colonization was to spread the Gospel and civilization. The *practical* drive was of course to make profit (pp. 317–20).

Although the nationalist stirring made itself felt in parts of Asia even before World War I – in India the first National Congress was held in 1885 and from 1910 Indians received voting rights for provincial assemblies – it was after the War that the era in which an emerging articulate, western educated, native elite, began to raise demands for self-determination. The early nationalist movements were mainly urban based, but World War I, led not only to the recruitment of thousands of Indian, North African Arab, and African black troops to serve in the French, and British armies, but also transported thousands of Chinese and Vietnamese laborers to Europe to work behind the front (pp. 320–323). Returning home the colonial soldiers and workers brought with them new aspirations. In a climate influenced by the ideas of the Mahatma Gandhi and by the Russian Revolution, India was launched on the way to independence by the Government of India Act (1936) which gave the vote to 35 million people and enabled the nationalist Congress Party to gain electoral victories in eight out of eleven provinces.

With hardly any exception, Asian and African states entered life after World War II, under slogans of modernization, which meant the expectation of higher standards of living, improved health care and educations etc. Many of these hopes were frustrated because of the absence of political stability and of an efficient administration. The only successful countries were those with a high level of ethnic homogeneity and educated elites – South Korea, Taiwan, Singapore and Japan. However, Japan was a case apart. Having not been colonized, and opened to the West since 1853, Japan turned itself into a modern state. By the mid-1870s it possessed a parliamentary-type government, independent courts, a functioning bureaucracy, and armed forces based on universal conscription, as well as an education system that very soon found its mission propagating a virulent form of nationalism and emperor-worship (pp. 323–324).[5]

During the fifty years since 1945 the total number of states has more than tripled. However, the great majority comprise countries which, until recently, were not states in the sense as states are defined by Professor van Creveld.

In the chapter entitled "The decline of the state: 1975– ", Professor van Creveld elaborates his reasons for believing why since the advent of nuclear weapons the state is in decline. In essence, he relates the decline of the state to five fundamental developments: the waning of the likelihood of major wars; the retreat of public welfare arrangements; the internationalization of technology;

the new types of threat to internal order; and the withdrawal of faith in the
state's administrators and administration. Taking the Hobbesian character of
mankind and society for granted, and believing that states only exist to prepare
for and to wage war, Professor van Creveld concludes that, though nuclear
weapons may provide deterrence, their use has put an end to the idea that self-
preservation can be attained by victory in war and has thereby destroyed the old
conception of the state's raison d'être. Given that the state arose out of the
success of a small number of "absolute" monarchs defeating universalism on
the one hand and particularism on the other and consolidating territorial
domains and concentrating political power in their own hands, and given that
in order to wield both civilian and military aspects of that power they set out
to construct an impersonal bureaucracy as well as the tax and information
infrastructure necessary for its support, Professor van Creveld regards the last
mentioned characteristics of the state as its hallmark. Therefore, deprived of its
old raison d'être, the state, about a century and a half after its birth, began to
provide itself with a new ethical contents (p. 415).

Professor van Creveld ignores the entire humanist tradition of western
society and the decades of labors' and liberals' struggle for social and
economic improvement and equity, and believes that deprived of the ability to
wage war it was to secure its survival that the state turned inward and
constructed the Welfare State. He is aware that "from about 1840 on socialist
ideas, translated into practice", had worked in the direction of "turning the state
inward" (but forgets that the New Deal predated World War II, and Churchill's
post-war elections declaration that Labor's programme was a "Thieves'
Charter") and presents the declared "freedom from want" in the "Atlantic
Charter" (1942) as a kind of gift by the states to the workers to compensate
them for their war effort (p. 354). In this context, Professor van Creveld
describes the Welfare State in a way that makes the reader think that next to
eating from the fruit of the tree of knowledge in the Garden of Eden, its
creation was mankind's second Original Sin. Surely, one cannot as he does,
reduce the lively humane democracy that followed the barbarism of the 1930s
and the war, to the "justification" of state's hunger for finding new reasons for
intervention, and dispose of it saying that "throughout the interwar period,
there existed a sprinkling of mostly middle-class, left-wing intellectuals in
many European countries who looked at developments originating in Moscow
as their shining example; feeling ashamed at what they called poverty amidst
plenty". (p. 356). Equally bizarre is his comment on Keynes and Keynesian
economics. Does Professor van Creveld really think that deficit financing
(which he calls "deficit funding") was really as he writes "an elegant name for
putting the printing presses into operation? (p. 357).[6] In one respect, however,

he is correct, namely that left- and right-wing ideologies apart, professional economists encouraged state intervention and a certain kind of Keynesian demand regulation (called the neoclassical synthesis) which led to state intervention to explode (pp. 356–357). Indeed, most of the nationalization which took place in the 1930s and 1940s had been carried out by left-wing governments for ideological reasons, and against opposition from the right and that even conservative cabinets found themselves powerless to resist them. Sometimes, Professor van Creveld writes, the need to provide employment acted as the decisive factor; in other cases it was a questions of enabling bankrupt companies to continue providing essential services in fields as far apart as transportation and defence (p. 358). In truth however, as anyone who lived through the period in which the Welfare State was created could have told him, its creation was not the bureaucracy's attempt to perpetuate its position but the expression of the will of millions of people who had seen the Great Depression and the war to establish a more humane world. It was precisely the old establishment and its bureaucrats that were trying their worst to bring back the pre-war social and economic relations. Moreover, the adoption of Keynes' economics, of which nationalization of industries was not an integral part, was not the product of nationalism, but the expression of a search for a better, more secure and mankind-friendly, world.

Though in most countries the pace of nationalization showed signs of slackening after 1975, a few witnessed a great expansion of the public sector during the late 1970s and even the early 1980s. In the United States, the world's richest society and most committed to free-enterprise capitalism, federal welfare schemes had made little progress since the days of the New Deal. However, in the late 1950s and early 1960s, there too was society shocked into action by a number of inquiries which revealed how the less fortunate Americans lived amidst all the plenty by which they were surrounded, and once the budget-minded Eisenhower had left office, reforms designed to correct this situation were started under the Kennedy administration. They were greatly accelerated by Lyndon Johnson who coined the term Great Society to describe them (pp. 358–360). Whatever their precise form, the various programs had been designed to assist weak population groups such as the elderly, the sick, and, later, single mothers; however, it soon turned out that the greater the benefits offered, the larger the number of those entitled (p. 362–363). And so, in both Europe and the United States – to say nothing of developing countries – the expansion of state-directed welfare led to an equally great growth of the bureaucracy. To support all these bureaucrats, the share of government spending as part of GNP grew to proportions which, except in periods of total war, were without precedent in history (p. 361).

While this is true, Professor van Creveld overlooks that it was not only the state bureaucracy that expanded but likewise the bureaucracy which administered the private sector.[7] He does not see that it is in the character of large-scale enterprises and multinationals to be as bureaucratic as states, that the shift toward a service economy is by its very nature stimulating a new style of bureaucratic feudalism, that state enterprises are sometimes more and sometimes less efficient than private enterprises and that their efficiency cannot always be measured by profit and loss accounts alone but by the service they provide for the society. After all, thousands of private enterprises go bankrupt every year which is hardly evidence for their superior efficiency, not to mention how the efficiency of British Rail has fared as soon as it became denationalized. More significantly however, Professor van Creveld ignores the entire cultural legacy of pre-war Capitalism which the supporters of a more humane world had to overcome, and that high social business, and academic, position gives access to television, radio and press, and therefore, the voice of economic advantage, being louder, is regularly mistaken for the voice of the masses.[8] In short, what he tells his readers in this context is the trite assertions of the voice of advantage, namely that "initially most of the governments confronted by these problems refused to look them in the face", but that by the end of the 1970s, "squeezed by a combination of rising taxes and inflation, and fearing a future which promised nothing but greater burdens, in one country after another the electorate signified its disgust with the welfare state and those who promoted it" (pp. 363–364).

It was not Milton Friedman[9] but Margaret Thatcher who became the most strident representative of the new economic conservative reaction and whose cure of "the British enema" provided a beacon for numerous other governments which followed along the same road in the 1980s (pp. 369–370). It is true that even communist China started rolling back state control over the economy. In 1978 the Party's new secretary-general, Deng Xiao-ping, formally announced the "Four Modernizations" (p. 373) but Marx was never concerned with the problems of economic growth. He thought the economic growth was the historical task of Capitalism. In his own words, "the bourgeoisie, during its rule of scarcely one hundred years, (in 1848) has created more massive and more colossal productive forces than have all the preceding generations together . . ." (Marx & Engels, 1848) and that it was later societies which were to distribute the fruits of this success more equitably. China has not renounced state control. It only attempts to avoid the mistakes into which Lenin has led by standing Marx on his head as Marx himself claimed to have done with Hegel's dialectics.

Professor van Creveld comes to the conclusion that with the renunciation of the Keynesian inspired economic theory and practice the attempt to give the state a new contents after it lost its ability to wage wars has failed. Toward the year 2000, economic policy in most countries, like the science of economics which provided it with both an explanation and a justification, had made a complete about-turn. The trend toward greater state intervention in the economy which started in the 1840s and gathered steam after 1900 or so was dead or dying; its place was taken by a renewed emphasis on private enterprise and competition (p. 376). Almost everywhere governments struggled to retain at least part of the welfare state, including above all elementary and secondary education, but that apart, the dream of using government to "lift up" the masses was clearly in ruins and indeed even avowed "left-wing" parties took a centrist stance and declared themselves no longer socialist. The old forms of political-economic organization have been largely discredited, and the search for others to take their place is on.

The other elements undermining the state and leading to its decline are the spreading of the new technologies and ecological problems technology created, and the expansion of trade, which compel governments to work together (pp. 377–388). But in most cases it was the multinationals, rather than states, which were the first to develop and deploy the most modern technologies (pp. 389–390). They need the state to provide stability and defense. The greater percentage of the assets belonging to the citizens of each state was likely to be located beyond its borders; and the vital economic decisions which affected such things as investment and employment inside each state were likely to be made by people over whom it had no control, and even the states' ability to control their own currencies was crippled (pp. 390–391). To be sure, governments did not lose all influence over currencies. They could still manipulate the money supply and retained control over some key interest rates such as the discount rate, but citizens and foreigners act with equal ease, ignoring their respective governments while moving money in and out of any given country by pressing a button. The unprecedented development of electronic information services seem to mark another step in the retreat of the state (pp. 391–392).

While the last years of the 20th century have not ushered in an age of hermetically sealed empires, Engsoc, and thought control, and obstacles to "globalization" remained formidable, and important states participate in a very large number of global, regional, or merely technical international organizations, and have retained only parts of their sovereignty, their control over both their economies and citizens' thoughts has undoubtedly declined (pp.

391–393). Even the provision of security, which in Professor van Creveld's opinion, has at least since Thomas Hobbes been recognized as the most important function of the corporation known as the state, is again increasingly shared out among other entities (p. 406). Thus the likelihood grows that the state will lose its monopoly over those forms of organized violence which still remain viable in the nuclear age and become an actor among many (pp. 406–407). Finally, the bureaucracy, that Hegel praised in 1830 as the "objective class" which put the public good above its own, and Max Weber regarded the embodiment of "goal oriented rationality", has today probably not an individual left who believes that such are its attributes. In fact, since the 1960s study after study revealed state bureaucracies to be self-serving, prone to lie in order to cover the blunders that they commit, arbitrary, capricious, impersonal, petty, inefficient, resistant to change, and heartless (p. 408).[10]

Having said all this Professor van Creveld arrives at the conclusion that at this time almost everywhere states are showing a declining willingness to take responsibility for their economies; to provide social benefits; to educate the young; and even to perform the elementary function of protecting their citizens against terrorism and crime. At the close of the second millennium, he says, the state is not so much served and admired as endured and tolerated (p. 414). However on the positive side, he assumes that states are much less likely to engage each other in major hostilities – let alone warfare on a global scale. "The devil's bargain that was struck in the 17th century, and in which the state offered its citizens much improved day-to-day security in return for their willingness to sacrifice themselves on its behalf if called upon, may be coming to an end" (p. 408).

In this reviewers opinion, what Professor van Creveld described in his final chapter as the decline of the state seems more the the symptoms of loss of direction than what he says to be peoples' "disgust" with the Welfare State (p. 364). It shows the end of the roaring wind of progress and original thinking called Modernism, and the end of the Modernist achievements from town planning to social security. It signals the beginning of an era in which people immerse themselves in their daily business and immediate circle, and turn their back on the political parties who until now claimed to command the levers of change, and a retreat to living materially from day to day, and intellectually from hand to mouth, in a community deprived of compassion, solidarity and the search for truth.

Professor van Creveld has written a very lucid elaboration of his views. His book covers an impressive length of time and geographic territory and should therefore make a most welcome contribution for students of political science.

Social and economic historians will be less enthusiastic because of the almost total absence from the narrative of the wider background of the processes he mentions. They will miss the analytical brilliance of Marc Bloch's *Feudal Society* (Bloch, 1953), the circumspection of R. H. Tawney's *Religion and the Rise of Capitalism* (Tawney, 1948) and M. Dobb's *Studies in the Rise of Capitalism* (Dobb, 1937), the transition to the modern state in science and technology and their cultural implications as discussed by J. Bronowski in *The Common Sense of Science* (Bronowski, 1951) and by many others, and they will miss the insight in 19th century political evolution one finds in books like Ch. Seignobus's *Histoire Politique de L'europe Contemporaire 1814–1896* (Seignobus, 1908).

NOTES

1. *Vide* Professor Warren J. Samuels "The Political-Economic Logic of World Governance". Conference paper for Tenth World Congress of Social Economics. Cambridge, England. August 2000.

2. For the analysis of this period Professor van Creveld rightly draws the attention of the reader to the work of F. L. Ganshof *Feudalism,* (Ganshof, 1961) and to M. Bloch's wonderful study *La Société Féodale* (Bloch, 1961).

3. J. L. Cohen. "Civil Society" in the *Concise Routledge Encyclopedia of Philosophy,* London, 2000, Routledge, pp.142–143.

4. *Vide* page 5, and note 3 for the difficulty with the interpretation one has to give to Professor van Creveld's use of the term "Civil Society".

5. Surprisingly Professor van Creveld does not note that like in Germany the industrialization of Japan started in the 1870s.

6. Having read as much as he did, it is really surprising that Professor van Creveld did not read either Keynes' *Treatise on Money* or *The General Theory of Employment, Interest and Money*, before postulating on the role of economists, and refusing to admit their share in making post-war society, at least until the 1980s, more humane than earlier societies.

7. See Y. S. Brenner and N. Golomb-Brenner (Brenner & Golomb, 1996).

8. See J. M. Galbraith "The Conservative Onslaught" (Galbraith, 1981).

9. Already in the thirties Milton Friedman was studying the feasibility of introducing an income maintenance scheme "negative income tax" by which households with an income which falls gelow some break-even level receive payments from the state related to the level of income, to alleviate poverty in New Jersey.

10. Surprisingly Professor van Creveld thinks that this is only characteristic of the state bureaucracy and not of the administrators of private enterprise as well. He does not consider the likely possibility that modern society is on the way back to some kind of social and economic feudalism akin to the one Europe experienced in Middle Ages.

REFERENCES

Benn, S. I. (1967). State. In: P. Edwards (Ed.), *The Encyclopedia of Philosphy*, Vol.VIII (pp. 6–10). New York, Macmillan Publishing Co. Inc. & Free Press.

Bloch, M. (1940). *La Société Féodale*. (*Feudal Society* translated into English in 1961 and published in London by Routledge & Kegan Paul Ltd.).

Brenner, Y. S., & Brenner-Golomb, N. (1996). *A Theory of Full Employment*, The new feudalism: managerial oligarchy. Boston. Kluwer Academic Publishers.

Bronowski, J. (1951). *The Common Sense of Science*. London, Heinemann.

Cohen, J. L. (2000). Civil Society. In: the *Concise Routledge Encyclopedia of Philosophy* (pp. 142–143). London: Routledge.

Dobb, M. H. (1946). *Studies in the Development of Capitalism*. London: Routledge and Kegan Paul, 1963.

Galbraith, J. K. (1981). The Conservative Onslaught. In: the *New York Review of Books*, January 22, 1981.

Ganshof, F. L. (1961). *Feudalism*. New York: Harper.

Hobbes, T. (1651). *Leviathan*. Oxford: Blackwell, 1946, Ch. 13.

Keynes, J. M. 1936. *The General Theory of Employment Interest and Money*. London: Macmillan, 1954.

Marx, K. & Engels, F. (1848). *The Communist Manifesto*. Works (MEW) Vol.4, Berlin, 1959, p. 5.

Sabine, G. H. State. In: E. R. A. Seligman (Ed.), *Encyclopedia of Social Sciences*, Vol.XIV, 1953 (pp. 328–332). New York: Macmillan Company.

Seignobus, Ch. (1908). *Histoire Politique de L'europe contemporaire 1814–1896*. Paris, Librairie Armand Colin.

Tawney, R. H. (1922). *Religion and the Rise of Capitalism* (Holland Memorial Lectures) Reprinted in 1948 by Penguin Books, West Drayton.

A PIED PIPER THEORY OF THE RISE AND FALL OF STATES

S. Todd Lowry

Martin van Creveld, *The Rise and Decline of the State*, New York: Cambridge University Press, 1999, Pp. viii, 439. $54.95.

The theory of the state has been a subject of formal abstract discussion over the past century or two. As the sources of power and their proper relationship to good government became subjects for policy, the concept of the state as an instrument of control emerged. Of course, Marx's definition of the capitalist state as an instrument for implementing the interests of the bourgeoisie has been the touchstone for much modern debate.

Martin van Creveld has written a unique exposition of the historical evolution of governance and the state. He focuses on the forms of political organization from tribe, city state, empire, and church without a clear analytical reference base except their statelessness. However, in the period from 1300–1648 in Europe, he formulates a vague theory of the state as the struggle by charismatic leaders to monopolize power or force. This evolution is played out in the form of struggles with the church, the feudal nobility, and the towns with the monarch emerging as the absolute authority. This constituted the nascent state. In order to implement and administer this centralized monopoly of power and military force, money had to be developed and extensively used. Also, a dependent bureaucracy was evolved to handle the collection and dispersal of revenues. Thomas Hobbes is identified as the first political scientist with a theory of the state.

Creveld comments briefly on the theoretical contributions of Erasmus and Machiavelli and the "Mirror for Princes" literature (pp. 183–184). It is unfortunate that more attention was not given to this generally ignored genre of administrative advice literature that dates back to antiquity. More attention and analysis should be directed to this tradition of administrative science that the author traces back to Xenophon's *Cyropaedia* (p. 172), a romanticized account of the education and training of Cyrus the Great. Creveld unfortunately identifies the subject of this work as Cyrus the Younger, the pretender to the Persian throne and leader of the military expedition into Asia Minor in which Xenophon participated. The question should be raised as to whether the existence of a formal tradition of statecraft correlated to individual authority for two millennia does not suggest that a broader definition of "the state" is in order. Otherwise, some analytical basis for distinguishing the administration of power in early modern times from that in antiquity would be expected.

What is disturbing is that the author proceeds to ramble through extensive histories of the European states from 1648 to 1789, tracing their political development, but with little reference to the economic and ideological growth of the commercial economy and ideological nationalism. The state is treated as an expression of an individual leadership function, monopolizing and administering military and economic power. These pedestrian histories are continued through the period 1789–1945 in which the power structure of the ideal state was perfected, dominating the people and the economy. Another collection of histories of various parts of the world, from Eastern Europe to Latin America, from 1696–1975, follows indicating that individuals were able to establish shells of states by carrying the image of the European state into areas of the world where they had little substance. This potential for a "Pied Piper" to play the tune of modern statehood in third-world countries and establish the authoritarian trappings of a modern pseudo-state is never adequately explained or analyzed.

It is all irrelevant, however, because, after the apotheosis of the European state during the period of the two world wars, it has begun a twofold decline. First, the extension of the authority of the state into protecting and augmenting the welfare of its citizens has been rejected and its rapid erosion has gained speed over the past thirty years. This process is exemplified by Thatcherism in the U.K. and Reaganism in the U.S. Privatization is negating public programs from public housing, military conscription, prisons, and regulation of business. At the same time, the emergence of giant corporations, whose international scope defies regulation by individual states, has made the state increasingly irrelevant in the emerging world economy. This is the ultimate conclusion of the author in appraising the felicitous evolution of the "Devil's bargain" that created the 17th-century state.

The author has clearly overextended himself in an attempt to generate a cosmic survey of human history keyed to a narrow political power-oriented concept of statehood and its contemporary dismantling at the hands of private enterprise. He has not systematically incorporated the serious literature on the theory of the state, or the development of the economic basis of the modern nation state. For example, the classic work by Franz Oppenheimer, *The State: Its History and Development Viewed Sociologically,* dating from 1914, is not even cited. This work follows the age-old Aristotelian tradition of devoting an introductory chapter to surveying the literature defining the subject matter under discussion. The author's work suffers from the vagueness and narrowness of his oblique definition of statehood. He would have done well to have consulted Herman M. Schwartz's book, *States Versus Markets: History, Geography, and the Development of the International Political Economy,* 1994. Schwartz frames his analysis in terms of the interaction of states with international market pressures, drawing on von Thünen and Ricardo for spatial perspective. His initial chapter, "From Street Gangs to Mafias", carries the same political thesis as Creveld in describing the emergence of the early modern state but is supported by citations to more secondary analysis. It offers an economic perspective based on von Thünen's actual experiment (p. 12) that demonstrated that a two-horse wagon loaded with grain with two men conducting it could travel only 220 miles on early 19th century European roads before the grain, the energy supply for humans and horses, was exhausted. This demonstrated that a round trip of 110 miles was an energy limit. It explains why grain was seldom transported more than 20 miles for sale in inland trade.

Only water-borne shipments were feasible which meant that most trade was international. Inland areas were self-sufficient micro economies and centralized authority required a monetary system to support activities and assert political hegemony on the mainland.

Schwartz's analysis of the evolution of the interaction between states and markets finds the U.S., the European Community, and Japan as the major states vying for control over world markets. He concludes that the U.S. is in a global position similar to that of Great Britain in the 1890s with the difference that the U.S. does not live under the shadow of the looming presence of an equivalent to the U.S. economy faced by the British.

Another recent work that parallels Creveld's is Graeme Donald Snooks's *The Ephemeral Civilization: Exploding the Myth of Social Evolution,* 1997. This work is built on the histories of ancient and modern societies and states. It is framed in terms of an analysis of the development of institutions in interaction with the economic and social challenges of the times. It concludes that institutional development is linear, not cyclical, and that new structures continually emerge. This line of analysis parallels Creveld's, but Snooks concludes that a limited number of mega-states will emerge and dominate the market adjustments in the world economy. These states will be the U.S., the European Union, China, and a revived Russia. A "fifth paradigm" is anticipated characterized by a "global state".

By comparison with these two works, Creveld's book is analytically challenged, and his historical elaborations lack focus and relevance. They are essentially excursions reviewing text-book political history. His primary conclusion – the significant thesis of the book – is that privatization and international corporations spell the decline, if not the demise, of state power. This is a linear projection of a very recent and short historical trend – no more than 30 to 40 years. It also cries out for some anticipatory analysis of the emergence of a new private center of power, the monopolization of force, that characterized the emergence of the state. In other words, what is the nature of the anticipatable international corporate mafia that will emerge. Is the Russian experience of privatization replacing much of the state's functions a prototype for the world economy? Or, on the contrary, is the private market system something that operates free from assertible monopolization of economic power?

Along with the problems raised above, one is concerned that Cambridge University Press's rigorous standards of review and editing are slipping. The book should have been refereed by someone with a better grasp of the literature in the field of international economic trade and history. A bibliography would have greatly assisted the reader in checking for references and added to the scholarly value of the book. References listed only at the bottom of the page greatly reduce the value of a work for research purposes. In addition, readers and copy editors should be expected to catch trivial errors such as the reference to tiger teeth as decorative elements among African chiefs, (p. 17; there are no tigers in Africa) and the reference to "Gustav Hayek" (p. 367) as the spokesman for the Austrian school of economics should have been caught. (Friedrich A. von Hayek is generally known to economists). Other historical pronouncements should have been questioned such as the assertion (p. 158) that troops in close formation armed with pikes spelled the end of the effectiveness of armored cavalry in the 14th century. The hoplite method of fighting dates back to the 6th or 7th century B. C. and most scholars credit the longbow, the cross bow and the musket with neutralizing the advantage of mounted knights over non-professional troops from the towns.

In general, while this book provides the reader with a galloping review of modern western political history, the concluding descriptive appraisal of the decline of the modern welfare state is not supported by the body of the work. It is an idiosyncratic projection of a short period of political and economic policy, namely, privatization. Its only supporting thesis is the general theme of the book that statehood and political organization result from the force of leadership by individuals that carry organizations with them, whether they are appropriate or not. This reduces to a "Pied Piper" theory of political history that may have some merit in the short term as illustrated by the careers of Margaret Thatcher and Ronald Reagan. However, such episodes may be aberrations that should be suspect as long-term historical theory.

REFERENCES

Oppenheimer, F. (1999). *The State: Its History and Development Viewed Sociologically*. New Brunswick and London: Transaction Publishers, reprinted from the 1914 Bobbs-Merrill edition.

Schwartz, H. M. (1994). *States Versus Markets: History, Geography, and the Development of the International Political Economy*, New York: St. Martin's Press.

Snooks, G. D. (1997). *The Ephemeral Civilization: Exploding the Myth of Social Evolution*, London and New York: Routledge.

VAN CREVALD MYTHHISTORICUS

Kenneth Mischel

It is Book Four of Thucydides' *Peloponnesian War*. The struggle between the Athenians and Spartans has entered its seventh year. An Athenian fleet bound for Sicily with instructions to first call on Corcyra is forced by inclement weather to dock on the island of Pylos instead. Demosthenes, who happens to be travelling with the fleet, is unable to convince the fleet's commanders, Eurymedon and Sophocles, of the island's strategic value. He doesn't have to. The troops, who happen to be looking to relieve their boredom, spontaneously set about the task of fortifying the place even though they lack the iron tools to do so efficiently. They have time to make headway in this task because the Spartans happen at the moment to be caught up in celebrating a festival.

When the Spartan response does come, Lady Fortune continues to smile upon the Athenians. When men and arms are particularly needed, a thirty-oared privateer belonging to the Messenians, allies of Demosthenes, just "happen[s] . . . to put in an appearance." And when the Athenian help finally does arrive, the Spartans somehow forget to block a crucial entrance to the harbor.

A conciliatory Spartan delegation approaches the Athenians. "It is not reasonable . . . for you to think that because of your present strength and your recent acquisitions, fortune will always be on your side. True wisdom is shown by those who make careful use of their advantages in the knowledge that things will change." The Athenians rebuff the Spartans' words and offer of peace. In doing so, they, like Agamemnon, the triumphant vanquisher of Troy, set themselves up for a reversal of fortune. Only a single year later, the balance of the war begins to tip in the Spartans' favor. The rest, as they say, is history. And for Thucydides, whom according to Francis MacDonald Cornford's classic *Thucydides Mythistoricus* wrote with one eye towards the war and the other towards the Greek stage, the rest is a good bit of Aeschylian tragedy as well.

Cornford's Thucydides has a contemporary kindred spirit in Martin van Crevald. The subject of the latter's sweeping and insightful historic tragedy *The Rise and Decline of the State* is not Athens but a roughly four-hundred year old organization of government originating in Europe and eventually making its way around virtually the entire globe. As it was for the Athenians and Agamemnon, so too for the state a reversal of fortune may be in order. The last part of the book is dedicated to showing that the early stages of such a reversal may already be at hand.

For van Crevald, what most distinguishes the modern state from other forms of government that came before it, save classical city-states, is the concept of publicness, that something can belong to an abstract entity with a persona of its own. Telemachus in search of his father can tell Menelaus that he has come "on his own business, not of the people," while a person serving as a juror can attend to a business that is not our – or any other flesh and blood person's – own. Such a clear distinction between public and private would have gone unrecognized by a pre-modern emperor or tribal chief. Evidently, it was unclear to Hapsburg Emperor Charles V, who in his 1543 testaments to his son continually precedes words such as "treasury," "servants," "army," and "countries" with the possessive "my."

How the sense of a public domain came to resurface in Europe in a fundamentally new form is the subject of the work's second and third chapters. In the first of these, van Crevald organizes a narrative of medieval monarchs' circuitous routes to increased power against the background of these monarch's struggles against the Church, Empire, nobility and self-governing towns. This allows the work of an impressive array of historians and some primary texts to be synthesized in a coherent, compelling way. Briefly, we observe a changing balance of power between church and monarch through the juxtaposition of Henry II of England's public repentance for the murder of Archbishop Thomas Beckett with the French kings' effective imprisonment of the papacy in France about a century-and-a-half later. Of the Empire, we glimpse the pull that the concept of a universal secular Christian organization had on Europeans, even as the concept reified was fading. We learn how the monarch's efforts against their imperial rival spawned some particularly interesting paths not immediately taken. For example, fifty years *before* the Peace of Westphalia and three-hundred-and-fifty before the UN, the duc de Sully, claiming to speak in the name of France's Henry IV, proposed abolishing feudal ties linking rulers across countries. In place of these ties, Europe was to be organized around a constellation of fifteen states, continuous in both geography and sovereignty.

If the monarchs in their struggles with the Church and Empire navigated against the top currents of pan-European universalism, they in their struggles with nobles and towns stayed a course against the undercurrents of particularism. "There was scarcely a king whose entire holdings were located inside his own realm (whatever that may have meant); scarcely a king, too, whose realm was not riddled with independent and semi-independent principalities of every sort." While the monarchs' universalist foes were eventually defeated, their particularist foes were gradually co-opted. Nobles were turned from competitors into "king's men" in return for (initially) retaining numerous privileges such as freedom from taxation and being granted a near-monopoly on the upper ranks of the monarchs' governments. Self-governing towns lost the power to self-govern but the burghers' loss of republican liberty was compensated for by increased economic empowerment, as for example by consolidating states' facilitation of national markets.

Was the triumph of the monarchs' inevitable? Was the Athenians' triumph at Pylos? "Had the Emperor also been the head of the established religion, as was the case in virtually any other part of the world where similar political systems existed, then almost certainly his power would have proved suffocating and the modern state would never have been born."

Having destroyed or co-opted their competitors, it might have seemed to early modern "absolute" monarchs that there were virtually no limits to their power. Yet an absolute monarch cannot govern his realm alone. Mechanisms to formulate, transmit and carry out his will are needed. And a funny thing happened over the course of a couple of centuries. As these mechanisms grew and became increasingly impersonal, they gradually weaned themselves off their monarchs, to the point that the latter were no longer needed for functioning. Spain's King Philip II once rebuked an ambassador for insisting upon some ceremonial protocol and was rebuked in turn, "Your Majesty is himself nothing but a ceremony."

The evolution of the state into a mature, by virtue of being abstract, entity is explored in the book's third chapter. The Roman Empire, which at its zenith counted between 50–80 million people, never had more than a few thousand centrally appointed bureaucrats. France, by contrast, had a population of between 18–20 million and a central bureaucracy of 25,000 in 1610 and about 50,000 in the early years of Louis XIV. As van Crevald points out, it is no coincidence that a Frenchman coined the term "bureaucracy."

Large tax revenues, which were needed to feed the French bureaucracy and an exponentially growing standing army as well, were made possible partly by the mass of personal information (births, deaths, marriages) that the state now began to collect. Indeed, as early as the 1670's, Louis' minister Colbert was arguing that the state, if anything, was collecting too much revenue in taxes. Still, much of this revenue remained in the hands of localities and nobles. The Revolution did away with this problem, as nobles were replaced by *intendants*. Governmental upheavals continued to come and go, but through them all the impersonal character of the French state continued to strengthen. The French model was either forcefully exported or partially emulated throughout the rest of Europe. A landscape of – to use a phrase of Balzac's cited by van Crevald – "giants wielded by pygmies" had emerged.

"And once he slipped his neck in the strap of Fate, his spirit veering black, impure, unholy, once he turned he stopped at nothing . . ." So tells the chorus of Agamemnon's sin. Even if his daughter Iphigeneia's death by sacrifice seems ordained, must Agamemnon comply with this outcome so zealously?

Must he tell his henchman, "Hoist her over the altar like a yearling; give it all your strength?" *The Rise and Decline of the State*'s fourth chapter is devoted to the ways in which these new giants – "liberal" as well as "totalitarian" – came to devote themselves to secular idolatry with an Agamenonian fury. "States demanded and obtained sacrifice on a scale which, had they been able to imagine it, would have made even the old Aztec gods blanch."

A tone of retrospective horror, much like that of the Greek chorus, "What comes next? I cannot see it, cannot say," punctuates the text. We learn of the power of romantic folk ruminations married to the pretensions of the state, exemplified by flags, anthems and self-celebrating holidays. We learn of some of the ways in which states' internal defense forces – in Britain between 1805–1842 the number of prosecutions (adjusted for growth in the population) rose by factor of 4.5 – and public education systems transformed their populations. We learn of the ways in which states "conquered" money, and the implicit tax powers it implies. Finally, we learn about total war. The Battle of Rocroi (1643) featured 48,000 troops; three decades later, the Battle of Malplaquet featured 200,000. If 18th century armies sought to conquer provinces, 19th century armies sought to conquer whole countries. As for 20th century-armed forces? "On the sixth of August 1945, a fine summer day, a single heavy bomber appeared over Hiroshima and dropped a single bomb. Moments later the sky was torn open. A thousand suns shone, 75,000 people lay dead or dying, and total war, which the states of this world had spent three centuries perfecting, abolished itself."

Like Agamemnon, whose inner yearnings to be a secular god cause his eventual undoing – it takes but a minute's conversation and Clytemnaestra has him treading the divine tapestries – van Crevald's 20th century state has itself partially to thank for planting the seeds of its likely reversal of fortune. Van Crevald starts with war. Total war's "perfection," culminating in nuclear armaments capable not only of devastating foes but the entire world, has to this point had the consequence of decreasing the likelihood of total war's renewal. Armed conflicts are becoming increasingly regional in nature. Yet, preparing for, and waging, war is a chief means by which states exert an emotional pull on their populations. If they cease doing so, "there will be no point in people remaining loyal to them any more than, for example, to General Motors or IBM." Which arguably is why Werner Siefert, the current CEO of the Frankfurt Stock Exchange and planned CEO of iX, the pan-European exchange combining the stock exchanges of Frankfurt and London, can be so nonchalant about his anticipated move to London. The point is also supported by the way in which the unacceptability of casualties has shaped the military responses of the U.S. and its European allies in recent operations in the Person Gulf, Bosnia and Serbia. The unacceptability of casualties has another consequence. As can be seen in Sierra Leone, mercenary forces are staging a return. Such forces had been sent home by newly sovereign post-Westphalian states three and a half centuries ago.

Another unintended, unforeseen consequence of the state's drive to hyperfunctionality has been the recent retreat of welfare. The welfare state has in some ways been the victim of its own success; while the number of U.S. children rose by 41% between 1952–1972, the number of such children eligible for benefits under the Aid for Dependent Children Program increased by 456% over the same span. Furthermore, the gobbling up of the "economic" by the "political" in the form of nationalization of industry has, on hindsight, been a failure. A common denominator among most industrial nationalization programs the world over is that they eventually went from black to red. Australian Prime Minister John Curtin once expressed the sentiment that "predominantly, government should be the agency whereby the masses should be lifted up." At the turn of the 21st century, those in need of being lifted up would be well advised to turn elsewhere.

A similar theme applies to the state's relationship with technology. Van Crevald notes that from the point of view of the state, the trajectory of technology is a Janus-faced creature. One face looks out on a region in which the technology at hand is used by the state to increase its sovereign scope by increasing the efficiency of its operations. The other face gazes upon a region in which the state's would-be competitors, among them intergovernmental organizations, non-governmental organizations and multinational corporations, use the technology to gain sovereignty at the state's expense. The need to globally coordinate the telegraph communications gave birth in 1865 to the International Telegraph Union and marked the first time states authored a fictional person authorized to make certain decisions binding on them. More recently, the European Economic Community has duplicated the feat. If things go as planned, in 2002 it will hold a monopoly on the sovereignty to issue official European currency.

Multinational corporations also represent fictional persons whose place in the world cannot easily be placed within the neat borders of a standard globe. In which domain would you circumscribe a Toyota manufactured in the U.S. by American citizens?

In the realm of Janus-faced technologies, van Crevald might have turned to the Internet. Initially the vestige of the United States Department of Defense, its packet switching capabilities designed with the cold war in mind, the Internet has often been turned into a vehicle for regulatory and legal "arbitrage." The encryption program Pretty Good Privacy (PGP) is used by numerous non-governmental organizations and groups communicating under repressive regimes. Despite being originally considered as a possible munition, and therefore subject to a U.S. ban on export, the code for PGP was put on the Internet in the early 1990s. A subsequent governmental investigation was eventually dropped. To the extent that they realize that they cannot police the Internet alone, the U.S. and other states may ultimately come to look upon the "private" use and development of encryption programs as beneficial to their own interests.

If the notion of publicness originated with the classical republics and city-states, what lines link modern political theory with its republican theoretic predecessors? Among the heroes of van Crevald's account of the former are Bodin, Hobbes, Locke and Montesquieu. But emphatically not Machiavelli, who is seen here primarily as the author of *The Prince*. But what about the Machiavelli of the *Discourses*? What about the thread that John Pocock, Quentin Skinner, Bernard Bailyn and Gordon Wood, among others, trace from the writers of classical Greece and Rome through those of *quattrocentro* Italy and early-modern England on into the conceptually earnest founders of the early American republic? The question is particularly important to the extent that van Crevald's thesis is generally correct, because addressing it may be crucial to willfully shaping the cloud of possible organizations of governance from which our future will emerge. As the Russian literary critic Mikhail Bakhtin once put it in his *Problems of Dostoevsky's Poetics*, a "genre is always the same and yet not the same, always old and new simultaneously." For this reason, Daniel Defoe was able to spot a continuity of the emerging European order with the wisdom of the ancients, despite the attacks of republican polemicists using the very same texts and some scruples of his own. Modern, state-based political theory springs forth from a profoundly rich genre. If we go back to the genre with an open mind (we don't want to invent that which is not organically there), new ways of understanding the ancients so as to order ourselves may well emerge.

Why is *The Rise and Decline of the State* an important book for economists to read? To the extent that its thesis has merit, some of the fundamental functions that states have performed in the past will surely be taken up by entities we have until recently unproblematically assumed to be purely "economic" in nature. In light of this transformation, and in light of Viner's dictum that economics is what economists do, just what economists do may turn out to be in store for a transformation of its own.

VAN CREVELD'S ANALYSIS OF THE STATE: A NOTE

Warren J. Samuels

Martin van Creveld's analysis of the rise and decline of the state is not only theoretically ingenious but perceptively organizes the vast history of the political organization of society. Much of his analysis, not surprisingly, turns on his definition of the state. Quite aside from his definition, however, there is much more to be said about political organization (e.g. the state). Inasmuch as I feel no need to restate a summary of van Creveld's analysis and interpretation, I shall move directly to a brief statement of what else needs, in my judgment, to be said.

Some of what I want to say involves a jump from van Creveld's focus on the state as an abstract entity to political organization (e.g. the state) as a decision-making structure and process (but without negating his historical sequence). Let me begin by recalling an exchange I once had with the late Murray Rothbard at a professional meeting, an exchange over the definition of the state. Rothbard wanted to substitute a host of ostensibly private entities for those of the modern state. These institutions would substitute for courts and armies, and so on. I argued that this new institutional apparatus would now comprise the state. He demurred.

Let me next turn to the distinction between government and governance. Government is the physical manifestation of what van Creveld calls the state; or, better, van Creveld's state is the idea of an abstract entity of government formed during the later stages of his sequence. Governance is the structure and process by which decisions are made that have important impact on other people. It is a structure and process that encompasses both what we call public or official government and what are nominally private, nongovernmental centers of decision making (including the firm, apropos of, e.g. Ronald Coase's and Gardiner C. Means's theories of the firm). A society run by an absolute emperor, by an East India Company, or by the Established Church would not have a recognizable government but would have a structure and process – a system – of governance. Similarly, in a society with both conventional government and viable private corporate enterprise, (1) both government and corporate sectors would comprise the system of governance and (2) they would not only interact but each would be instrumental in the development of the other.

The first point I want to make is that however the state (or political organization) is conceptualized, a structure and process of decision making does exist. The most basic decisions, as it were, pertain to the structure itself and its operation. This process therefore determines whose interests are (1) to be incorporated in the structure itself and (2) to be promoted by that structure; in sum, whose interests are to count – with due regard to the dependency, at least in part, of the perception and identification of interests upon the structure itself. That is to say, the process determines whether Alpha's or Beta's interest will prevail when they are in conflict in a particular sphere of action.

This process correlatively also governs the formation of the basic economic institutions, whether they are the feudal manor, the trading company, the modern corporation, the labor union, property, negotiable instruments, etc., such as they may exist.

This process also involves the choice of conflicting sets of customs and/or would-be customs, e.g. between different groups of businessmen, between business and labor, between entrepreneurs and suppliers of capital, and/or between business and consumers.

All this is important for both the economic organization of society and determining actual economic performance.

Also involved in this – and amply manifest in van Creveld's attention to the social production/control of a compliant and docile labor force – is what Bertrand de Jouvenel wrote was common to all social systems, to wit, the control and use of the human labor force.

One can, therefore, either superimpose upon or substitute for van Creveld's analysis a further consideration: the extent to which government/governance is narrowly controlled or widely diffused. One can – I do not say must – argue that one important aspect of the history so neatly categorized by van Creveld, perhaps the important aspect, involves the widening range of interests/groups with access to and participation in decision making. This pluralism has taken one or another form of democracy. This is at least as important as the emergence of the idea of the state as an abstract entity and all that it involves; indeed, the emergence of the idea of the state as an abstract entity may only be the rhetorical means of comprehending and advancing democratization: in order to escape from the equation of the ruler with the state ("I am the state" or even "I am the only president you have"), the idea of the state as an abstract and separate entity is formulated. I personally remain open-minded or ambivalent as to whether the overall concentration of power has decreased, increased, or remained essentially the same over the centuries.

But surely there is a difference between an aggrandizing bureaucracy and the attempts by the hitherto disadvantaged to use the state in the same manner as the hitherto – and, truth be known, continued – advantaged had/have been doing, namely, to promote their interests as they understand them. Van Creveld's approach either constitutes or lends itself to a defense of past uses of the state, e.g. in forming economic institutions and in determining whose interests are to count. This is done, in part, by reifying property and other rights and by implicitly treating them as part of the natural order of things, whereas the so-called Welfare State is treated as an artificial contrivance. But such is an ironic matter, inasmuch as van Creveld both acknowledges and indeed insists upon stages in the development of political organization – stages that involve changes, however gradual and subtle, in the structure of rights and economic institutions.

Which brings me to the further question: What is the relation between ideas – including the ideas of the divine right of kings, of the state as an abstract entity apart from king or temporary office-holder, and so on – and material structures and relations? Max Weber's total system had a place for both the dominant system of belief or philosophy of life – as expressed in religion – and structures of power. Causation, it seems to me, runs both ways in a system of cumulative causation, dialectics, general interdependence, or overdetermination: ideas influence structure and structure influences ideas. Van Creveld seems to emphasize the primacy of ideas over structure, but that begs the questions of the origin and basis of acceptability of ideas. Would a Thomas Hobbes or a John Locke or an Adam Smith have written what they did at any other time, and would their ideas have been so influential?

Further topics also arise: First, the distribution of power in the non-governmental sectors of governance. Second, the distribution of power between nation states in a world in which decisions by one state influence the opportunity set of others. Third, the fate of democracy and pluralism, political and economic, in an increasingly corporate world economy.

The critical result of van Creveld's analysis is to put the modern state in historical and sociological perspective. First, the modern state is seen to be but one stage in the evolution of the political (=decision making) structure of society. Second, following and generalizing Adam Smith, it is true both that government exists to protect property – the rich from the poor, the large from the small owner – and that property (in whatever system it, like government, exists) is what it is in part as a function of government. Third, that both government and property are driven by social structure – and also influence social structure. And fourth, that the critical concept is not government but governance – inclusive of nominally official or public government and nominally private decision making, the latter especially concentrated private-sector decision making. This phrasing is largely due to the language of the stage of the modern state but can readily be applied, *mutatis mutandis*, to the other stages.

Finally, I advance the foregoing with due regard to the fact that different definitions – such as of the "state" – tend to embody different theories of that which is being defined.

Multiple Review of Morgan and Morrison's
REVIEW OF MODELS AS MEDIATOR: PERSPECTIVES ON NATURAL AND SOCIAL SCIENCE

MEDIATING MODELS

Ross B. Emmett

Cambridge and New York: Cambridge University Press, 1999. 401 pp., xi, 0-521-65097-6.

Economics today is dominated by mathematical modelling. That has not always been the case, and one of the roles historians of economic thought have adopted is that of translating earlier scientific activity into the language of models to give contemporary currency to older ways of doing science. But modern economists model, and it is therefore appropriate to spend some time reflecting on the question of what a model is, and what scientific role it plays. For the historian of economics, this exercise can also shed light on the question of how one explains the development and use of economic models among modern economists.

Mary Morgan and Margaret Morrison have provided us with a closer examination of these questions by asking contributors to this collection of essays to look at models in economics and the physical sciences. Most of the papers included were presented at one of several forums organized by the

Research in the History of Economic Thought and Methodology, Volume 19A, pages 143–159.
Copyright © 2001 by Elsevier Science B.V.
All rights of reproduction in any form reserved.
ISBN: 0-7623-0703-X

editors and Nancy Cartwright at the London School of Economics, the Wissenschaftskolleg zu Berlin, the University of Amsterdam, and the Tinbergen Institute. The papers included in the volume provide a good mixture of case studies examining economic models and models from physics and chemistry. Economic models treated in the volume include: Marx's Schema of Reproduction (Gert Reuten); the national econometric models used by the Dutch Central Bureau of Statistics (Adrienne van den Bogaard); Irving Fisher's modelling of the quantity theory of money (Mary Morgan); and the business cycle models of Frisch, Kalecki, and Lucas (Marcel Boumans). Physical science models examined include: the Ising model of critical point phenomena (R. I. G. Hughes); quantum models (Cartwright); the MIT-Bag and Nambu-Jona-Lasinio models in hadron physics (Stephan Hartmann); and various chemical models (Ursula Klein). The volume also includes two philosophical treatments of the role of models in science, by Morrison and Mauricio Suárez, and a general introduction to the issues by the editors.

If asked, economists would probably say that their models are representations in testable form of the theoretical relationships among economic variables. That is, the model in some fashion *stands between* the relationships of economic theory and the data of the real economic world. If a model "works" it provides evidence that the theoretical relationship is true. And, if the theoretical relationship is upheld, the model may provide a means to control future events in the real world. Some economists (those who have read Deirdre McCloskey) might add that a model helps to tell a good story about the economic variables.

Were we to probe a little deeper, the economist might suggest a relationship between model, theory and the real world that implies the model's *dependence* upon either (or both) theory and the empirical data, as in Fig. 1. According to this view, the model is governed by the theory – real world axis. (some might

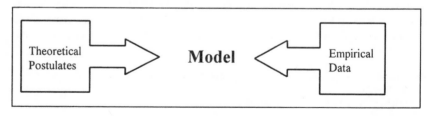

Fig. 1. Model Dependence.

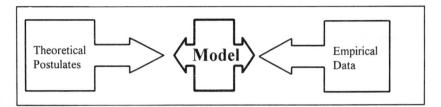

Fig. 2. Model Independence.

want to draw the arrows from theory to model to the data, and others might want to draw the arrows the other way, but in either case, the model's dependent position is emphasized; see p. 93 for a similar diagram provided by Boumans.) The economist would probably argue that the model's dependent position is crucial to its scientific role: logically, the economist's use of the model to either explain the data or refute some aspect of the theory is only effective if the model *does not introduce something new* into the space between theory and empirical reality. Nancy Cartwright and Margaret Morrison confirm that physics shares this same assumption about the dependence of models.

But what if the model is not completely dependent on the data or theory? What if it does introduce something new into that space between theory and empirical reality? The essays in *Models as Mediators* suggest, and the editors argue in their contributions, that scientific models *do introduce something new* into the space between theory and empirical reality, and, hence, that they are independent elements in the scientific process. Figure 2 represents their argument, indicating that, while the model stands between the postulates of theory and the data of empirical reality, it is independent of the two, and hence contributes *something of its own* to the scientific process.

So, if the model is not dependent on either theory or the data, how should we understand it? What role does it play? The authors of these essays provide a variety of perspectives on the independent role of models. One common metaphor is that of a caricature. Hughes quotes Michael Fisher: "a good model is like a good caricature: it should emphasize those features which are the most important and should downplay the inessential features" (p. 128). And Reuten quotes Allan Gibbard and Hal Varian, who suggest economic models "give an impression" of economic reality and involve "deliberate distortion . . . in a way that illuminates certain aspects of that reality" (p. 198).

The notion of models as caricatures introduces the economist as caricaturist. Here we begin to see the import of the volume's philosophical analysis for the work of a historian of economics. If we can accept the analogy of a caricature, then an examination of the development of economic models must incorporate an account, not only of various models' logical differences and progressions, but also of the *selection of attributes to feature* by the economists involved. Differences among economic models may be attributable therefore not only to theoretical differences, but also to other aspects of the economist's life and work: the mathematical concepts or techniques known, her public policy perspective, and the range of metaphors or analogies available.

An excellent example of such a story is provided in the essay by Marcel Boumans. Boumans examines the development of three different macro-economic models: Kalecki's Marxist-inspired business-cycle model; Frisch's Rocking Horse Model; and Lucas' general equilibrium business-cycle model. Boumans shows that each model was constructed from different ingredients, and that the mixture of these ingredients was specifically chosen by the economist. His conclusion provides an even stronger statement of the models' relationships to theory and empirical data than that represented in Fig. 2 above. Figure 2 shares with Fig. 1 an implicitly linear relationship between theory – model – empirical data. Boumans places the model at the center of a multi-dimensional space (represented in Fig. 3 below as a circle, but more appropriately thought of as a sphere), including not only theory and empirical

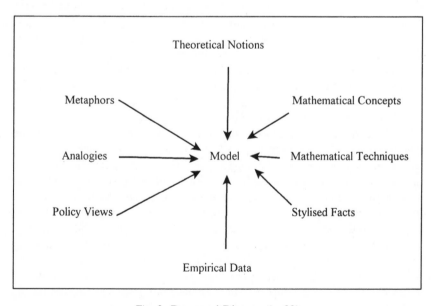

Fig. 3. Boumans' Diagram (p. 93).

data, but also mathematical techniques, public policy perspectives, the stylised facts accepted by the researcher, and the metaphors and analogies available to the researcher at the time.

The notion of a model as caricature highlights the role of the modeller as caricaturist because it interprets the model as a representation. Most of the essays in the volume deal, in one way or another, with this issue of representation: what is it that the model represents and how does it represent it? But this focus on representation, as interesting as it is, obscures another aspect of models which is discussed frequently in the essays –the model as a *technology*. Morgan and Morrison speak of models as tools (pp. 32–35) and Klein tells us that chemical formulas were conceived as "paper-tools" for the building of new chemical models (p. 137). Focusing on the model-as-technology does not mean that the model is not representational. Instead, it draws to our attention the uses to which economists put models, for technologies are developed to change or control things. So too, models "function as a technology that allows us to explore, build and apply theories, to structure and make measurements, and to make things work in the world" (p. 32).

Bogaard's essay on the development of the Dutch national accounts provides a nice transition from the model-as-representation to model-as-technology metaphor. The goal of the national accounts is to represent what is happening in the Dutch economy. But in order to understand the development of the Dutch model, Bogaard argues, one must also see it as a "social and political device . . . a practice connecting data, index numbers, national accounts, equations, institutes, trained personnel, laws, and policy-making" (p. 283). Prior to the 1930s, initial efforts at measuring the economy concentrated on the development of an economic barometer which would indicate when the economy was rising or falling. This work followed efforts at Harvard to establish a similar barometer for the U.S.A. The failure of the barometer to predict the depression of the early 1930s led to a new research program, focused on building a model of the Dutch business cycle. Bogaard shows that the decision to move from barometer to business-cycle model and the model's subsequent development, were guided less by theoretical issues than by decisions regarding: the choice of personnel with different training (Jan Tinbergen's appointment in particular); the opportunity to use new mathematical techniques; political demands for more accurate prediction; the organizational structure of Dutch public life in the political bureaucracies, universities and institutes; and the need to acquire better data. The development of the Dutch economic model and its national accounts, then, is guided by a larger set of issues, and Bogaard illustrates the use of the new model-as-technology not only as a scientific instrument, but also as a political tool to gather data and guide policy-making and a academic tool to determine the future of disciplines and university activity.

> The technology was some sort of 'liberation' both from the uncertainties caused by the whimsical nature of the economy and the woolly theories of the economists. At the same time the technology defined both new tasks: how to make and maintain the model and how to use it; and new roles: roles for economists, roles for politicians, roles for statisticians and roles for universities (p. 323).

Mary Morgan's contributions to the volume take the model-as-technology metaphor one step further. For Morgan, a technology is not only a tool which enables you to control reality, for in the process of using the tool, you also learn something about reality. The model is both a means to knowledge and a source of knowledge (p. 35). Using Irving Fisher's models of the quantity theory of money as her case study, Morgan shows that an economist learns both from building a model and from using it.

> Learning from building involves finding out what will fit together and will work to represent certain aspects of the theory or the world or both. . . . Learning from using models is dependent on the extent to which we can transfer the things we learn from manipulating our models to either our theory or to the real world (p. 386).

Learning from models and transferring the knowledge gained to either theoretical work or our participation in the real world leads us to the metaphor which gives the book its name: the model as a mediator. Mediation is an appropriate term to use here, because it captures the notion that the models do not constitute the relationship between theory and empirical reality, but rather mediate between the two. For the economist, the notion of a model as a mediator brings to mind banks, which are described as intermediating savings and borrowing. The bank example may help us understand the mediating role attributed here to models. Banks are not passive channels through which savings and loans flow: their existence structures the financial flows, provides criteria for eligibility and contribution, facilitates the creation of new financial flows, and so forth. In a similar fashion, models in Morgan and Morrison's perspective structure the scientific interaction between theory and empirical data, facilitate the development of new theory and data, and provide criteria by which the scientist can explain and control.

We have learned to see the possibilities created by the independence of financial intermediaries in the market process, and perhaps we can now learn to see how model independence functions in the scientific process. Morgan and Morrison have done philosophers and historians of science a service by providing such a close examination of the mediating role of autonomous models.

WHAT ARE MODELS FOR?

Stephen Thomas Ziliak

Learning how to do things with models makes up the bulk of graduate education in the contemporary school of economics. A Web page calls out: "Do you like to build models"? And replies, "then economics is for you". Graduates go on to universities or to non-academic employments to spend their days building, manipulating, testing, adapting, or abandoning their models of the economy. It is – they have reason to believe – what they get paid for.

Doing things with models has become central to economic inquiry. When someone in the seminar asks "Where is your model"? she is pointing at once to a tacit hierarchy of knowledge production and to the police squad that is standing at its factory gates. Yet practitioners have put little to print on the art and philosophy of modeling. It is a learning-by-doing skill, with some oral history attached. Not long ago a philosophically and historically-minded student listened to an economist complain about "the over-lapping generations model." The student (who was new to economics) thought he was complaining about dialogic interpretation in a period when the Keynesians were getting older and the New Classicals were getting younger. ("When the Berlin Wall fell, did it move to Minnesota?") In other words, neither have historians or philosophers written much about models and their place in a scientific economics. With the aid of a few simplifying assumptions (the articulation of which I leave to the reader), the situation can be modeled with a two character, one act dialogue.

> *Graduate student:* How do you build a model?
> *Professor:* Build one.
> *Graduate student:* But *how*?
> *Professor:* Hmm.
> *Graduate student:* Well, what relation shall my model have with theory and data?
> *Professor:* Good question!

No one doubts that they *are* good questions, but no one has bothered to answer them in a systematic fashion. Perhaps, in the end, an explicit answer is not possible. To date the answer has been like that of the parent to her child on a two-wheeler: "we can't tell you *how* – just do it, and we'll tell you when you are doing it well." Obviously, many are doing it well.

Competence with models – models that are seen to be 'useful' – is important. It is unusual to publish a paper in an economics journal that does not refer to a model built by someone else. More likely, the published paper describes a model of some recent fashion; it manipulates that model until the author has found a version that (for current purposes) is seen to be superior, and (once properly equipped with an urn-model of probability) it then carries out an estimation of statistical proxies of that model. Still, the scientific status of models is a matter of some controversy. Paul Feyerabend once said that a science going through a period of axiomatization is a science that is going to be dead. You don't have to be Paul Feyerabend to wonder about a science that is going through a period of model-manipulation, as economics is.

To kill the silence (not the Feyerabend question) Mary S. Morgan and Margaret Morrison offer a curata in their *Models as Mediators: Perspectives on Natural and Social Science* (1999). A collection of essays written by philosophers and historians of economics, chemistry, physics, and mechanics, *Models as Mediators* attempts to frame actual practices of modeling, from building to abandonment, in the larger conversations of the philosophy of science. Keen to objections, the editors aim to unify their thinking about models across the social and natural sciences. Their argument proceeds by cases, taken mainly from the book. The cases that will be of interest to this audience are essays on the modeling strategies of Irving Fisher, Ragnar Frisch, Jan Tinbergen, Michal Kalecki, and Karl Marx.

Employing the case studies as illustration, Morgan and Morrison argue that philosophers of science would do well to think of models as "mediators" between theory and data. Useful models, they say, are "partially independent" of both theory and data (p. 17). It is the quality of independence, they say, that allows models to work as "instruments" or "tools" for the scientist exploring either theory or data; it is the quality of independence that allows the model-using scientist to learn things about theory and data – and to mediate the conflict between them (pp. 18–19, 32–36). Models, say Morgan and Morrison, are now a "foundation" for a "variety of decision making" (p. 8). The mood is celebratory. Their investigations suggest finally that "models should no longer be treated as subordinate to theory and data in the production of knowledge . . . No longer", they say, "should (models) be seen as 'preliminary theories' in physics, nor as a sign of the impossibility of making economics a 'proper' science" (p. 36).

* * *

Models can be expressed as an equation or as a set of equations; they can be expressed as a graph, a flow chart, or a Venn diagram; models may be deterministic or stochastic; they may take the form of a three-dimensional object (like a molecular structure) or they may take the form of a well-turned metaphor (like a poem by Adrienne Rich: "I gave my tongue to love, and this makes it difficult to speak"). Whatever the outward expression of a model, a model is constituted by something. A model may be constituted by an analogy. Pairing familiar objects in unfamiliar ways (like sex and censorship or children and refrigerators), the analogy often seeks to spark new thinking – instrumentally – about something we thought we knew, in theory or in fact. The philosophers Max Black (1962) and Mary Hesse (1966) have made a persuasive case for the idea that models are metaphors and analogies, doing just that.

Like a model Swiss train is to a real Swiss train, a model may be constituted by a representation of a real object or system the model seeks to describe, matching part with part, relation with relation. The tight match of a model with a real system it seeks to describe is the property of "isomorphism" admired by some economists and more skeptics.

Models have other constitutions. Models may be constituted by "simulacrum" (following the work of Nancy Cartwright) or by "nonlinguistic entities" (the semantic view of Tarski, van Fraasen, and Giere). But in choosing an analogy to govern their inquiry – "models as mediators" – Morgan and Morrison signal their strong attachment to the Black and Hesse view of models. Of course the idea that an *economic* model should be seen as metaphor and analogy is not novel. It was suggested by Veblen long before models had any currency, and the idea circulates in contemporary economics through an influential body of work by Deirdre McCloskey (1983, 1985 [1998], 1991). The idea of a "toolkit," a cliche of textbook economics ever since Joan Robinson spoke of it, is now gaining philosophical support. (Morgan uses the word once in her essay, though without comment: p. 350.) But neither are Morgan and Morrison the first philosophers to propose that a model is useful precisely for its function as a "tool," like hammer is to nail, or thermometer is to pressure, a technology of knowledge production. H. S. Gordon had made the same proposal in 1991 in his *History and Philosophy of Social Science* (Gordon 1991, p. 108). Yet *Models as Mediators* does emerge novel in its endorsement of model "independence" and "mediation," and of course especially in its case studies of actual modeling practices.

The case studies range in quality from informative droning (p. 78) to dazzling social study of science (Adrienne van den Bogaard's essay on Jan Tinbergen is especially fine [Chap. 10] and so is Marcel Boumans' critique of conventional stories about Model, Theory, and Data [pp. 91–95]); variably, each case study will have staying power in future scholarship. I should hope that instructors will put van den Bogaard's essay on Tinbergen and Mary Morgan's essay on Irving Fisher (Chap. 12) into the hands of the macroeconomist who believes that models are disembodied sausage machines (the recent deliberation of the American Economic Association concerning the eugenics of Irving Fisher is more evidence of the truth that they are not; the failure of the *AEA* to consider during its deliberations the eugenics of Richard T. Ely is still more.)

But there are problems with the governing metaphor of the book. Take for example Morgan and Morrison's claim that models are "partially independent" of theory and data. And the corollary: that independence is necessary for models to do their work. The claims are dreamy – they get one puffed up with promise for an objective science. But they are easily enough deflated by actual practice, as for example in the history and economics of "welfare" in the United States.

From colonial times to the 1910s poor relief in cash and in-kind was everywhere a municipal affair. (A significant exception is the veteran's pension.) During the last quarter of the nineteenth century America's leading philosophers and providers of relief to the poor were members of the Charity Organization Society. Inspired as much by the Social Gospel Movement as by the social philosophy of Herbert Spencer, the Charity Organization Society believed that material assistance – cash and in-kind – could, when properly given, have a *positive* effect on labor supply and self-reliance. Suffice it to say that "proper" relief was defined by "scientific charity" (their term), a complex process of house-to-house fact gathering, the establishment of case histories, the maintenance of personal attention, the elevation of moral counsel, the active cooperation with other relief societies, and the "scientific" deliberation by executive committee (Watson, 1922; Ziliak, 1996, 1997). In other words, a dollar of assistance given scientifically could, they believed, *increase* the work efforts of the recipient, and build character. Neoclassical economists of the late twentieth century believe that the baseline model of work disincentives – a trade-off made by utility-maximizers between commensurable units of consumption and leisure – has been worked out satisfactorily. To the neoclassical economist in the contemporary scene, the idea of the Society is seen to be quaint and, in a foul mood, ass-backwards. De gustibus non est disputandum.

There is more where that came from. The labor market model of the Charity Organization Society could in principle generate the Society data, and a case can be made that it did (Ziliak, 1996, 1997). It seems that the Society data were "hypothesis-" or "model-laden" in no small degree (Gordon, 1991, p. 605; Ziliak, 1997). How so? In most cities the Charity Organization Society refused to relieve anyone of bad character (the intemperate, the adulterous, the shiftless, and the vicious – these were the bad characters, the "unworthy poor"). Thus the "unworthy poor" would not be observed in the Society's data because the data on the "unworthy poor" were not traced beyond the initial interview. The only data in the Society's books would represent the life-histories of the "worthy" poor – those who needed just a little hand up and who, *ex ante*, were most likely to be self-reliant. Of course relative to the "worthy" poor who were plagued with illness or injury, the able-bodied "worthy" poor, both men and women, would exit the relief rolls quickly, and more often with higher wages. Thus the Charity Organization Society was able to observe in the data the predicted effect of its model: "scientific charity" increases labor supply and builds character. Or preserves it, anyway.

H. S. Gordon has argued that the use of hypothesis-laden data is simply "*bad* science" (p. 605; his emphasis). We need not agree, by contraries, with the philosopher N. R. Hanson that hypothesis-laden data are inherent to science. But if we accept Gordon's characterization – which is not so far away, I think, from that of Morgan and Morrison – then we are obliged to explain why prominent social scientists of the day were active participants and thinkers of the Charity Organization Society. In England one can name John Ruskin, Henry Sidgwick, William Beveridge, R. H. Tawney, and Beatrice Webb. In the United States one can name Richard T. Ely, the Stanford economist Amos Warner, the Chicago sociologist Charles Henderson, the economist Edward T. Devine, and Mary Richmond (a founder of modern social work). In other words, Morgan and Morrison's requirement that a good instrument be "independent" (like hammer is to nail or thermometer is to pressure) is challenged by actual practice, the practice of the Charity Organization Society, which designed and administrated welfare programs for more than 30 years, and which attracted some of the most prominent social scientists of the day. It would be easy to supply other examples. Adrienne van den Bogaard supplies an example in her careful study of Tinbergen, the Dutch Central Planning Bureau, and the (Dutch) Central Bureau of Statistics (Chap. 10, pp. 322–323). To raise the point is not to justify the practice. But it does ask historians and philosophers of social science to keep in mind the difficult – and yet influential – cases of model- or hypothesis-laden science.

The appearance and the persistence of model-laden data deep inside the practice of a real-world institution asks philosophers to consider differently what models are for. I think it a mistake to believe with Morgan and Morrison that models are primarily "tools" or "instruments" for learning about worlds or theories. Unlike Gordon, Morgan and Morrison do bend toward the idea that models can also serve a "representative function" (p. 25) – isomorphism. They do not want their view confused with a classical instrumentalist view. But for Morgan and Morrison, the function of models stops here.

Models, like theory and data, I am trying to say, seem to play a meaning-making function that is not adequately captured by Morgan and Morrsion's explanation. A good mediating instrument, a good model, may be the one that is most useful in telling a scientist's story; it may be a model that mediates a sociological narrative with layers of epistemological and moral narrative meanings. It may be that the scientist's story, or rather her "authorial control" of the plot-lines (as they say in Theatre Studies), may not be negotiable – not in the short run. It is a problem of identity politics. But that is not all. Consider again the model-laden data of the Charity Organization Society. It was not in the constitution of the Charity Organization Society to allow some other agency to collect their data, interview their clients, provide moral counsel, and so forth. To do so would be to split the Society at the root – and at the cost of their principles. Likewise, it was not in their constitution to allow their clients to enter a control group – say, the relief rolls of the Township Trustee and his vote-buying ward boss – to better "test" the principles of scientific charity. Unprincipled relief-giving would make the Society's job of character building even more difficult. To put it another way, the meaning of their model was constituted by the data and by the process of data collection that the model generated, and the meaning of the data and their means of production was constituted by the model.

Here we find leading scientists of the nineteenth century living not in a meaningless tautology but in a meaningful story. This is a point that is anxiously hiding in the corners of the van den Bogaard and Boumans papers but can't find its way out (for example, p. 322). We often bring to phenomena, as the philosopher and cognitive psychologist Jerome Bruner points out, a story well worked out (Bruner, 1986; Winslade & Monk, 2000, pp. 52–53). We construct intentions, create situations, select strategies of procedure, participate in institutions, and deliberate plot-lines in a "landscape of action" which, Bruner has observed, already have "meaning" in story form. Our practices (in science or in law or in personal relationships) are then "performances of meaning." In the technology of our models we do not want to forgo authorial control over our performances. And yet the stories that may be told with our models are in many senses already worked out – in master metaphors of the speech community, in institutional incentives, in prior moral commitments, and so forth. A model is an instrument and a representative function, no doubt. But as Deirdre McCloskey puts it, in a similar vein, the model supplies something besides instrument and representation – she says models are an "answer" to a story (1991, Chap. 1). In the cases of the Charity Organization Society and of

Tinbergen's research program, the model assists a plot line and a network of meanings much larger than any of its parts, model or data or theory of survival. Morgan and Morrison do not devote much space to an explication of "mediations." As it stands I do not believe that they have coined a persuasive analogy for models. But I do think that with some elaboration of a notion of mediation as narrative, at once a ritual performance in a well-defined "landscape of action" and a competition for authorial control, they could make a go with the analogy "models as mediators."

Of course there is never just one true story, in society or even in the mind of most scientists. Scientists, like lawyers and others in a human society, are mediators of stories. The point of thinking in terms of "narrative mediation" in philosophies of the social context of modeling is similar to the point of thinking of narrative mediation in family law: it encourages the "teasing out of these stories in order to open up possibilities for alternative stories to gain an audience" (Winslade & Monk, 2000, p. 53). In this reading what is an important mediating function of models? It is to keep the stories of a scientist and his enabling institutions meaningful in a web of inter-dependent meanings – sociological, epistemological, and statistical. The macroeconomic models of Jan Tinbergen and the models of "scientific charity" endorsed by Henry Sidgwick and Richard T. Ely are more than mediators of theory and data. The inter-personal and institutional mediations of their models are ubiquitous. There are more than three-dimensions to the problem of mediation. Saying so helps us understand what Boumans and van den Bogaard and Reuten (on Marx) are up to.

* * *

Stripping themselves of a social context, Morgan and Morrison seem to believe that there are no "rules" for model making (p. 31). Without too much effort, however, one can employ the ideas of their contributors to the task of identifying two important rules. The first is the Rule of Precedence: your model had better be strongly connected to the models currently in fashion in your field. Reasons and kinds of connections will vary. But nowadays a model of economic growth had better square-up to the models of Robert Solow and Paul Romer. A model of unemployment insurance had better confront the job search approach of Nicholas Kiefer and the institutional approach of Tony Atkinson. The Rule of Precedence directs attention to a sociological point about narratives of cooperation, too often ignored in the Morgan and Morrison collection. It is the point that models seek isomorphism with *extant* models, the models that are believed to be authoritative and cutting-edge. A good model may have internal parts and relations that have a one-to-one correspondence with the system the model seeks to explain. That some philosophers desire isomorphism with a system is undeniable. But in economics the practice of seeking isomorphism with extant *models* is much more common, a practice of legitimation. It is the practice of saying that so and so (authoritative economist) assumed agents are Butter Side Up when an objective observer knows it would be better to assume they are Butter Side Down. It is the practice of letting me show you how my model is general and that it can test the power of Butter Side Up . . .

The Rule of Precedence can be illustrated with an example from the economics of labor market transitions. Economists have borrowed from industrial engineers and biologists various models of "survival." These models generate 1/0 outcomes with a time-ordered random variate as dependent variable. Examples are the life in hours of a lightbulb, the years of a human life following the growth of a cancer, and the number of months one stays on a welfare program. One of the key indicators in this class of model is the "hazard rate" – the probability of a "failure" at time t given that the subject has survived to time $t–1$. A technological problem with survival models is that the early exits – the early "failures" – have a relatively high probability of failure. This has the effect of lowering the average rate of failure, or raising the average rate of survival. The problem is that the scientist cannot use the survival model alone to separate independent effects (the heterogeneous characteristics of the subjects) from time-dependent failures. In the study of welfare programs, for example, one wants to know to what extent the long spells of participation (or "survival") are experienced by people with severe obstacles to self-sufficiency and to what extent the long spells represent a dependency *created* by the long period of time spent on welfare. The technological problem is seen to be analogous in the study of light-bulbs and cancer patients.

The Rule of Precedence in the study of welfare programs is to cite a short paper on "heterogeneity" and "time-dependence" by James Heckman (1991), estimate an exponential and a weibull regression model, and maybe plot the error terms, all of which shall (quite imperfectly) give some indication of the presence of time-dependence. The Rule of Precedence in survival models of welfare does encourage isomorphism – with James Heckman and the G. E. lightbulb. The point is that there are tools and evidence beyond the technology of the survival model that can throw light on the situation, perhaps more clearly. In the case of welfare, the rich field notes of caseworkers, anthropologists, sociologists, and psychologists can go a long way toward identifying "heterogeneous" obstacles to self-sufficiency. Yet their evidence is typically spurned by publications in economics journals. A similar observation could be made concerning health care professionals, medical journals, and their treatment of cancer patients. In a different vein, one sees the great cost of the Rule of Precedence in the economic studies of racial discrimination: despite the observation of Albert Hirschman and innumerable culture critics that racism is a "value," even economists of liberal or left persuasion continue to assume with Gary Becker that race is a 1/0 variable and that firms have "tastes" for discrimination. In models such as these the radical shift in the twentieth century of female black domestic servants into the clerical professions – indeed, any observed difference in black/white labor market outcomes – can be seen only from the perspective of firm behavior and employment law; the shift of blacks into clerical jobs or elsewhere cannot be seen in the Becker model as the outcome of blacks "performing" race – a putting on of the work face and voice, though in part, we know, it is. The Rule of Precedence in model-building hesitates to accept understandings so obtained.

The second rule of modeling is the Rule of Containment: your model had better be contained entirely (with no peep holes) by the master metaphor of your speech community (utility maximization or Marxian class struggle) or your model will not be "useful." It is the stronger rule. It cannot be violated, a lesson that graduate students learn the hard way (Klamer & Colander 1991). The Rule of Containment is a conserving force, a drag on the economic imagination, and is probably wasteful in the neoclassical sense of that word. Ready examples of mainstream exclusivity and wastefulness are to be found in the concerns of heterodox economists: in *Feminist Economics*, *Journal of Post-Keynesian Economics*, *Review of Radical Political Economy*, *Review of Social Economy*, *Journal of Economic Issues*, and a host of others. The editors of *Models as Mediators* have not seriously considered these points, the Rules of Precedence and of Containment, or whatever one might call them, and that suggests the real world relevance of the mediator metaphor.

Finally, it should be said that Morgan and Morrison rank highly any model that "performs" well but they do not say what their standards of performance are (p. 34). Presumably the ideas of how to build a model will connect with the ideas of how to build a model that performs well. Yet the authors never say. (They do say what they do *not* mean: isomorphism with a real system.) In places (p. 34) they seem to defer to some standard errors of actual practice, such as the usual resort to statistical significance. Deirdre McCloskey and I show in a 1996 paper in *The Journal of Economic Literature* that a conventional measure of "good performance" for a statistical model – statistical significance – should not be used as such. It is. A better measure of performance, we say, is economic significance (McCloskey & Ziliak 1996). What would the charity organizers say?

* * *

Models as Mediators is a welcome addition to the small but expanding scholarship that attempts to situate "ideas in context." The case studies are so full of information that they are bursting out of the editors' three-dimensional metaphor. That metaphor needs revising. A productive way forward would be to marry the researches of the London and Cambridge and Toronto Groups with the rapidly expanding scholarship in the United States on the "rhetoric of economics" – the new economic criticism. The place of models in a scientific economics is being examined with the tools of literary theory, ethnography, social history, and the sociology of science (Mirowski, Ed., 1994; Woodmansee & Osteen, Eds, 1999; Garnett, Jr., Ed., 1999; Ziliak, Ed., 2001). Morgan and Morrison's conclusion that "models should no longer be treated as subordinate" seems in its context – I am not sure they are aware – an imprimatur for the status quo of everyday neoclassicism and of what McCloskey calls a "search through the hyperspace of conceivable assumptions" (McCloskey, 1994, pp. 137–145). This is a story that more than one critic is going to agree with, and the meaning is already being performed.

REFERENCES

Black, M. (1962). *Models and Metaphors*. Ithaca, NY: Cornell University Press.
Bruner, J. (1986). *Actual Minds, Possible Worlds*. Cambridge: Harvard.
Gordon, H. S. (1991). *The History and Philosophy of Social Science*. London: Routledge.
Heckman, J. (1991). Identifying the Hand of the Past: Distinguishing State Dependence from Heterogeneity. *American Economic Review* (May), 75–79.
Hesse, M. (1966). *Models and Analogies in Science*. Notre Dame, IN: University of Notre Dame Press.
Katz, M. B. (1986). *In the Shadow of the Poorhouse*. New York: Basic Books.
Klamer, A., & Colander, D. (1991). *The Making of an Economist*. Boulder: Westview Press.
McCloskey, D. N. (1983). The Rhetoric of Economics. *Journal of Economic Literature, 31* (June), 482–517.
McCloskey, D. N. (1985 [1998]). *The Rhetoric of Economics*. Madison: University of Wisconsin Press. Second Edition.
McCloskey, D. N. (1991). *If You're So Smart*. Chicago: University of Chicago Press.
McCloskey, D. N. (1994). *Knowledge and Persuasion in Economics*. Cambridge: Cambridge University Press.
McCloskey, D. N., & Ziliak, S. T. (1996). The Standard Error of Regressions. *Journal of Economic Literature, 34* (March), 97–114.
Mirowski, P. (Ed.) (1994). *Natural Images in Economic Thought*. Cambridge: Cambridge University Press.
Morgan, M. S., & Morrison, M. (1999). *Models as Mediators: Perspectives on Natural and Social Science*. Cambridge: Cambridge University Press.
Watson, F. D. (1922). *The Charity Organization Movement in the United States*. New York: MacMillan.

Winslade, J., & Monk, G. (2000). *Narrative Mediation*. San Francisco: Jossey-Bass Publishers.

Woodmansee, M., & Osteen, M. (Eds) (1999). *The New Economic Criticism*. London: Routledge.

Ziliak, S. T. (1996). The End of Welfare and the Contradiction of Compassion. *The Independent Review*, *I*(1), Spring, 55–73.

Ziliak, S. T. (1997). Kicking the Malthusian Vice: Lessons from the Abolition of 'Welfare' in the Nineteenth Century. *Quarterly Review of Economics and Finance*, *37*(2), Summer, 449–468.

Ziliak, S. T. (Ed.) (2001). *Measurement and Meaning in Economics: The Essential Deirdre McCloskey*. Edward Elgar, forthcoming, Economists of the Twentieth Century Series.

Angresano's THE POLITICAL ECONOMY OF GUNNAR MYRDAL

POLITICAL ECONOMY OF THE DISASTER

Soltan Dzarasov

INTRODUCTION

Failure of market reforms, at least in the majority of CEE countries, is one of the most noticeable developments of our time. Their real outcome is in sharp contrast with the pledges and perspectives given at the beginning. Then, the President of the Russian Federation, Boris Yeltsin, giving the reasons for the necessity of market reforms to Parliament (the then Supreme Soviet) outlined an optimistic vision of their expected results in the form of quick and favorable changes in the lives of ordinary Russian citizens. At the end of 1991, he solemnly and confidently declared that after the six to eight month period of "shock therapy" which had to be gone through, economic recovery would be enabled to start and people's living standards start to improve. Encouraged by the authorities and the influx of capital, the mass media and other officials all said the same thing.

The problem of making yet another great historical shift in the life of an entire people was portrayed only in a simplified way, as if Paradise was waiting around the corner with the IMF and its advisors acting as guides on the way.

One needed, the theory went, only to break with the command economy, liberate it on the basis of "laissez-faire" principles, privatize state property, and

Research in the History of Economic Thought and Methodology, Volume 19A, pages 161–187.

a market would arise which, following the Western example, would lead us to prosperity and welfare with its "invisible hand".

With the help of this alluring theory, the authorities received a mandate from society to undertake the changes inherently necessary in following the neoclassical IMF model.

This deserves a few comments:

The IMF model is based on the foundations of the so-called "Washington consensus", that is, the basic principles recommended as a major framework for economic policy in the Third World and developed by the leading financiers and international financial institutions all mainly based in Washington. The American economist, Paul Krugman, summarized the consensus as "the belief that Victorian virtue in economic policy – free market and sound money – is the key to economic development. Liberalize trade, privatize state enterprises, balance the budget, peg the exchange rate, and one will have laid the foundations for economic takeoff" (Krugman, 1995, p. 29).

The greatest criticism of the consensus comes from economists who do not adhere to the neoclassical approach. For an observation of the results of the neoclassical approach, promoted by both the IMF and the World Bank in developing countries, see a work written by the Canadian professor, Michel Chossudovsky, with the meaningful title *The Globalization of Poverty: impact of IMF and World Bank Reforms*. It is noted in that article that "a new interventionist framework", sponsored by the Bretton Wood institutions, "has been engineered by macroeconomic policy, leading to the dismantling of state institutions, the tearing-down of economic borders and the impoverishment of millions of people" (Chossudovsky 1997, p. 15).

But these complications were not mentioned by Russian adherents of market reform, obsessed with genuine "capitalist zeal". A seasoned observer of transformation collisions, the Russian historian Dmitry Furman, has noticed a puzzling paradox in modern Russia. Under the guise of "democratic ideology", we experience a revival of "reversed Marxism":

> the real choice in the age of early Perestroyka, was probably between "Marxist reformation", set on preserving and even enhancing its symbols and tradition but outwardly rejecting a Marxist mythological scheme – this is the "tale with a happy end" where the hero reaches the land flowing with milk and honey and obtains the magic gift (a knowledge of the true form of property) and "reversed Marxism" which, on the contrary, rejects the symbols but preserves the mythological scheme. And I think that the first way would sooner lead us to "normal" democratic society with a free economy and to the society more "western" in type (Furman, 1998, pp. 92–93).

Meanwhile, this market tale is very similar to the communist tale which maintained that once private property was replaced by public property, then the source of welfare would flow to society. This communist theory was reinforced just at the moment in history when its opponent had been discredited – namely, the experience of the West during the 1930s when the free market lay in ruins. But the important distinction is that though Communist theory did not win the confidence of developed capitalist countries, for several decades it did enable the U.S.SR to obtain higher rates of growth than Western countries and significantly improved living standards. Nothing like this can be associated with the current market reforms.

Other CEE countries have not had so much to boast of. But the situation is not the same everywhere. Some of them did enjoy a certain level of growth although none of them ever achieved targeted levels of output and standards of living. It would not be an exaggeration to say that present market reforms have not justified people's hopes, which was the justification for the implementation. But this is not easily recognized. The advocates of market change stubbornly maintain the relevance of recent reforms – they do not see beyond immediate material interest and fail to acknowledge that certain sectors of the population, which gained from the conversion of wealth to public property have lost more than they have gained. Last year James Angresano published *The Political Economy of Gunnar Myrdal* in the U.S.A. Although Myrdal's name is widely known in the world for economic and social literature, to the reader from the U.S.S.R. Myrdal is known mainly by his two books published in Russian: *An International Economy: Problems and Prospects* (Myrdal, 1958) and *Asian Drama* (Myrdal, 1972). In his introduction to the former book, Myrdal posited that "basically the framework of my economic analysis differs from Marxist political economy and consequently it will face disagreement" (reverse translation).

In reality, Myrdal's books and ideas in general were favorably accepted in the U.S.S.R and provoked great interest, despite some differences of opinion. The Soviet reader was impressed with the humanistic spirit of the author's approach to the solution of the world's problems. Certain doubts had arisen among the Soviet specialists about the way to achieve independence and overcome poverty by the gradual integration of the colonial peoples in the world capitalist system. It has to be remembered that these books were written during colonial times. Marxists had never imagined a time of independence for colonial people and their gradual absorption into the capitalist system. That is why they formulated an alternative non-capitalist way of development for the third World countries.

The present great transformations which have taken place at the end of the twentieth century have completely confused all previous assumptions and perceptions. The Socialist alternative to Capitalism constructed on the pattern in the U.S.S.R. had failed. At the same time, only a few of the new free countries enjoyed the fruits of progress. The integration of these newly-free countries into the world capitalist system, involving the vast majority of the world's population, had not helped them to overcome backwardness in economic development, as Myrdal has dreamed, but had, on the contrary, made them poorer. Moreover, the majority of the people of the former socialist system had experienced deindustrialization, de-modernization and, as a result, had experienced increasing backwardness in social and economic terms. Both Marxist and Neoclassical market approaches appear to be models of development too narrow to reflect the ambiguous and contradictory processes of the emerging social and economic situation. Historical development has produced mixed forms of traditional and modern elements, which is quite close to the model developed and envisaged by Myrdal.

Specialists in CEE countries had already modified and implemented his approach which can be seen as an alternative to the Neoclassical concept. In consequence, James Angresano's volume, which illuminates the alternative approaches to real market change in CEE countries, is of particular interest at the moment.

We have become accustomed to the fact that solutions to all transition problems are seen by the specialists, all of whom come from the Western socioeconomic tradition, in terms of implementing the Neoclassical paradigm. So it is gratifying to hear a voice which is more in tune with the principles of free market reform developed by Myrdal. Angresano has brought together critical evaluations by other Western authors of CEE countries reforms looking at performances and outcomes. Much of this evaluation is new to us in the U.S.S.R., although such critical assessments are beginning to appear. Witness, for instance, the collective work, *Reform Perspectives by American and Russian Specialists*, which was published in 1996 in Moscow and contains a negative analysis of the results of transformations. Despite the authority of this work, and there are three Nobel Prize winners among the authors, Angresano's work represents a more fundamental and comprehensive explanation of the causes of the CEE reforms failure and provides the foundation of an appropriate alternative.

At the same time, it has to be said that if the book was translated into Russian, it would be perceived differently. Some would be interested, others would be disappointed and would try to disprove its significance. From the latter group of people – those who support radical change – it is possible to anticipate their automatic response. If the critic is Western, then he is declared to be incompetent in judging CEE reforms; if he is Russian, then he is labelled a conservative. However, the main achievement of Angresano's book is that the issues discussed are set out not only with great competence, but with unusual objectivity (unusual to what we are accustomed to from the West when analyzing CEE countries.)

In this review essay, I try to follow a certain logic.

First, I concentrate on the problem of the narrowness of the neoclassical approach and contrast it with a wider approach, both presented in the book as theories as well as relying on performance indicators as well. Then I consider the situation in CEE countries and China, including the situation in the field of economic education. Finally, I suggest some critical assessments of the advantages and omissions from the book as well. Given the exposition of the work, I prefer to cite the text in order to provide the reader with the possibility of working out his/her own opinion.

1. CRITIQUES OF NEOCLASSICAL NARROWNESS

Among the leading lights of socioeconomic thought in the current century, Gunnar Myrdal is particularly marked in taking a broad perspective of interests and types of activity. He was not only an economist but also a sociologist; not only a researcher, but also a teacher; not only a theoretician but also a practical politician, working hard to implement his ideas in real life. A good deal of Angresano's book consists of elucidation of these numerous activities by Myrdal, covering the evolution of his views. The pages devoted to the formulation of Myrdal's perspectives, showing one of the most prominent advocates of the institutional approach during a period of more than seven decades, will be read with great interest by all those interested in the work of this great thinker.

Myrdal's work is of great appeal to us in Russia because it contains a broad approach to social issues which we find applicable to our own processes and views we do not find elsewhere. Myrdal's special appeal is that he does not treat economic problems as isolated from other aspects of social life, which is a common feature of the Neoclassical approach, but he views them in close association with their setting and their determinants. Myrdal himself explained the essence of the institutionalism approach as follows: "the most fundamental thought that holds institutional economists together is our recognition that even if we focus attention on specific economic problems, our study must take account of the entire social system, including everything else of importance for what comes to happen in the economic field" (Angresano, 1997, p. 84). Such a perspective, therefore, sees the social system as comprising of economic and non-economic factors.

According to Myrdal, only a holistic approach, which he calls "institutional", is capable of revealing the inner character of the evolution processes and therefore could be called "logically tenable". Another central message of his institutional perspective he explains as meaning "history, politics, theories and ideologies, economic structure and level, social stratification, agriculture, agriculture and industry, population and development, health and education", as well as other ingredients in the social system which are in constant interaction. It follows that one cannot study social processes without taking into account the interaction of all the other factors.

Myrdal's institutional concept, as shown by Angresano, provides us today not only with academic value, but also with a practical means by which to consider the transformation processes occurring at the moment in the CEE countries and China. Although Angresano's book provides a fairly full exposition of Myrdal's views, it by no means just lists them. The author has undertaken his own comprehensive exploration of the situation of these countries. This double perspective – an analysis from wide interdisciplinary positions coupled with practical research – gives the book particular value as fundamentally developed theoretical research.

To quote Angresano: "To broaden his analysis of societies, Myrdal worked with scholars from other social science disciplines. He recognized that only a comprehensive interdisciplinary vision could identify the causes lying behind the pressing transformation issues. He emphasized the importance of not analyzing societal problems as isolated economic facts, but to evaluate them in their dynamic relationship with other aspects (such as cultural, historical, political) of society. Having assumed this heterodox position, Myrdal developed a method of analysis which avoided being doctrinaire and included a common-sense, open-minded approach to social problems" (Angresano, 1997, p. 120).

Using Myrdal's method of analysis and applying it to the countries undergoing transformation processes based on neoclassical orthodoxy, Angresano shows the pernicious results of such an approach. Myrdal in his time derided neoclassical economists for reducing complicated and ambiguous socio-economic processes to the level of optimum allocation of resources, and also for the belief that efficiency and favorable economic growth would be possible once the competitive market was established. The trouble with these economists, according to his opinion, was in an oversimplification of the reality which led to a firm anticipation that economic prosperity would follow the market. The author writes: "This vision viewed development as occurring in a linear, mechanical and simplified manner according to which neoclassical economists assumed that once certain economic conditions had been established (such as privatization of previously State-owned enterprises or a liberalization of the rules pertaining to free trade) the market mechanism would emerge and prosperity would inevitably ensue" (Angresano, 1997, p. 95).

This observation of the CEE countries in the book is so precise that nothing needs to be added. Unfortunately, in Russia, this simplified version of the complicated processes has been so often repeated from high authority, and every Russian citizen was regaled with same view by the media. Even such a way of thinking penetrated into academic circles; many economists accepted that the Anglo-Saxon Neoclassical model could be easily established in the conditions prevailing in Russia at the time. They were lulled by the promise of an economic panacea inherent in the model, while ignoring existing conditions. On this point, Angresano comments: "That neoclassical economists advanced their theories as universal propositions seeking to create 'an absolute and timeless economic theory' valid for every time, every place and every culture was 'foolish', according to Myrdal" (Angresano, 1997, p. 96).

Rejecting these universal doctrines, Myrdal insisted on a comprehensive exploration of every contradictory process. Something inappropriate in one case might be quite relevant in another. He applied this approach to the elucidation of the state role in the economy. He asserted that even in developed countries, the state was an essential determinant of economic recovery. As far as the Asian countries are concerned (and this was of particular interest to Myrdal), they adopted effective state-guided industrial and trade policies, having found themselves unable to conduct their policies without emphasizing the state role. An active role for the state is virtually necessary for the CEE countries as well, especially now to counter the "wild capitalism" and ensure lawfulness and social guarantees. Using Myrdal, the author notes: "The state is also necessary, particularly in CEE countries, to safeguard the quality of goods and services on offer. The extensive network of organized crime prevalent today in CEE illustrates the pressing need for greater state regulation of economic activity. The state is also necessary for providing a social safety net to its citizens who have come to expect social insurance and welfare schemes – much as they are provided by Western nations. State activity in this regard will become more necessary as privatization of previously state-owned enterprises continues and new owners have neither the desire or willingness to provide similar schemes for their workers" (Angresano, 1997, p. 173).

Here two important arguments are applied in favor of the role played by the state in the CEE countries' transformation. First, it is the pressing need for greater regulation of the economy. Second, it is the fact that the owners are obviously reluctant to meet the workers' need of social safety. However, we have to add that the state itself in the current conditions in Russia ("the state we're in", British economist Will Hutton would say) is an important source of criminalization in society.

A favourable setting for the rapid growth of mass organized crime was created by privatization, which enabled many entrepreneurs to make enormous fortunes "buying" huge state enterprises for negligible prices, if corrupt state bureaucracy allowed. As a result, unimaginable criminal liaisons embracing all spheres of business and the state emerged – liaisons unknown in neoclassical orthodoxy. The resulting situation is that the state cannot police criminality, being too often a part of it. This may seem an exaggeration to those outside the situation, but it is so serious that tomorrow it is possible that the world will see the transformation of a communist-bureaucratic totalitarianism into a criminally capitalist authoritarian society.

2. CRITERIA AND PERFORMANCE INDICATORS

The method of combining socioeconomic development criteria with corresponding indicators and applied to the situation in the CEE countries which is adopted in the book is of particular importance. Basing this criteria on Myrdal's own critique, the author separated the criteria of qualitative socioeconomic development as shown in the examples given from merely quantitative performance indicators (pp. 13–14).

This linking of the two sides of the social phenomena allowed Myrdal (and subsequently Angresano) to overcome the narrowness of the neoclassical approach. The latter school of thought, which is responsible for the CEE reforms, limits its analysis to the purely economic side, neglecting the peculiarity of nations – their historical and cultural traditions, some partially coinciding with each other and others distinctly different from each other. It is difficult to reflect with standard formulae and schedules all the variety, though within limits these are useful and sometimes necessary. But the tectonic shift from plan to market has such a deep and comprehensive effect on society that only the appropriate analysis of institutional transformations can set an adequate framework for the useful applications of certain neoclassical tools. The latter analysis is not therefore rejected – it must just be pointed out that only the qualitative approach makes quantitative methods relevant.

The book provides the system and criteria of socioeconomic development which overcomes neoclassical narrowness and gives much more adequate tools for an interdisciplinary approach. These tools are used for a comparative analysis of the outcomes of reforms in CEE countries and in China. Before discussing these comparisons, we should note the usefulness of the method, its unquestionable advantage compared to current practice whereby extremely different and variable processes are measured with the same, often irrelevant, indicators. The author himself explains his method as follows:

> Four steps need to be followed when evaluating and comparing the performance of selected economies: First: the selection of criteria which, taken together, comprise the analyst's definition of performance; secondly, the identification of performance indicators for each criterion; third, measurement of these performance indicators; finally, compilation of a performance index. This index consists of the quantitative performance measures of each criterion weighted according to its relative importance. This step can be taken if performance indicators lend themselves to quantification (Angresano, 1997, p. 11).

With the help of such a method, the author managed to present more reliable results from his comparative analysis than we find in other works. Whilst appreciating the content of the book in general, however, I would like to indicate that some criteria and indicators used are not very certain. Some really do characterize the socioeconomic situation, such as the criterion of equalization. The corresponding economic indicator is the distribution of wealth and income, the characteristic of a welfare state. It is difficult to produce criteria such as productivity and living standards which are precise and certain, particularly because an entire system of statistically measured variables could be used as performance indicators.

But this could hardly be said of a number of other criteria: (a) social discipline, rational planning, (b) national independence, (c) national consolidation, and (d) political democracy. The corresponding performance indicators are not so certain as the first ones. For example, the criterion of national independence is matched – according to the author – with political stability and ability to legislate, and the criterion of national consolidation with the ability to enforce laws. As I perceive it, the author himself oversimplifies. What does the ability to enforce laws mean? Should one consider such an ability as a performance indicator of national consolidation? In our view, scarcely ever. Authoritarian regimes are often as capable of implementing their law not less strictly than democratic ones. But as historical experience shows, particularly in the CEE countries, their consolidation was an illusion masking discontent and inner schism. At the first opportunity, in the social upheaval, it manifested itself.

Despite some skepticism of some of the solutions suggested by the author, his attempt to develop appropriate criteria and indicators for making a serious analysis should be supported and proven. He could be castigated for the scarcity of statistical data on which to support a criterion, but the framework of the analysis is the foundation of his work. But in my view such a rebuke would not be deserved. Today, the problem is not that we are lacking in data, but lacking in unbiased theoretical observations which can produce a complete analysis logically to the end. Following, we attempt to show that Angresano did not succeed in all cases. But where he did, he contributed a great deal. The most important thing is that he does try to present to the reader an objective picture and an honest assessment of the situation in CEE countries and in China using the Myrdalian model. This task determines his rigorous selection of the statistical evidence. This is justified because the main point for him is not the quantity of data but its reliability and its relevance.

Probably one observation should be made here. It concerns the lack of evidence in the book of the socialist period of development of the CEE countries. Some comparison should have been made in what occurred prior to 1989 and what happened later concerning rates of growth, levels of accumulation and investment, productivity and wages, income levels and living standards of the major part of the population. One of the positive results of the changes, it is noted in the book, is "the elimination of the necessity of shopping on the black market" (Angresano, 1997, p. 20) and that in Hungary "it is estimated that purchases on the black market account for up to one quarter of the average family's total spending" (Angresano, 1997, p. 179). This is quite true. But the main question is, did the reverse trends take place? Do they prevail? What does a comparative analysis of pre–1989 and post–1989 data say? Unfortunately, there is nothing of this in the book.

3. EASTERN EUROPE'S DISASTER

The central message permeating the book is the comparison of the opposing results achieved in their approach to market reform by China and the CEE countries. In both cases, we have a market transformation. But each chose different value foundations and development strategies. The book gives an eloquent comparison of the results of the China experience against the results in Poland, Hungary, Bulgaria and Russia. Particular attention is focused on a comparison of China and Poland, the latter selected as achieving the best results among CEE countries. Therefore, a pragmatic approach in one case (China) is opposed to an ideological in the other (Poland).

As can be seen from published data, while Poland achieved 5% annual rates of GNP growth only in recent years (1992–1996), probably a partial compensation for the sharp decline of 1991–1992, China enjoyed steady high rates of growth over a long period. Growth of more than 9% started from the first year of reforms in 1978 and continued consistently, at times rising to 14% and even more. These figures were characterized by a growth in productivity; the source of the expansion of Chinese exports, currency returns and the strengthening of the yuan. In comparison with the rapid growth of the Chinese economy, the data reflecting the decline and devastation of the Russian economy looks depressing. The reason is, says the author, that Eastern Europeans, including Russians, had chosen (or, as I think, were pressed to accept) an ideology and practice corresponding to the West which was less than compatible with their own interests. The book demonstrates with firm certainty that ideology remains an important ingredient of policy formulation and evaluation of performance, and, as a tool to promote the shock therapy view, is a basic foundation of the transformation policy.

The essence and targets of the changes were formulated in the CEE countries according to such an ideological bias: "Transformation goals for CEE in the early 1990s were biased toward Western interests, ridden with inherent ideological assumptions in neoclassical theory such as the desire to secure markets for Western exports and access to CEE's raw materials, and strengthen political influence throughout the region" (Angresano, 1997, p. 28).

The number of doctrinaire assumptions of the theory, illustrating the simplified approach to socioeconomic phenomena, are set out in detail in the book. According to the author they could be reduced to four basic points: first: strict dichotomy: either planned or market economy; second: transformation is perceived as a relatively easy process, reduced to liberalization and privatization; third: human behavior is characterized as being always fairly rational and in essence represents the pursuit of benefits. fourth: economic development is identified as the achievement of some sort of steady-state equilibrium, rather than as an evolutionary process in the social and historical context (Angresano, 1997, pp. 138–139). Ideological narrowness and simplification reveal themselves here clearly. All the various parts of the transformation process, embracing a complex of not only economic but also socio-cultural and political phenomena is reduced to the four assumptions mentioned above. If it is so simple then it follows that it is only natural to suggest that a direct cavalry charge will easily overthrow the system of planning and in its place the free market economy will instantly emerge. Commenting on such an approach the author writes: "The ideological, narrow perspective led to the application of microeconomic theory rules for economising to broad transformation issues in a simplistic manner. One recurrent theme from those partaking in the 'cavalry charge' to CEE was that rapid privatization was the prerequisite for the establishment of a 'free market economy' " (Angresano, 1997, p. 136).

This simplified vision of the transformation process encouraged many Western advisors to come to Russia. Attracted by beneficial contracts, they energetically set to work, having no idea about the complicated structures of the planned economy, nor about the historical traditions and mentality of the Russian people. As a rule they represented consulting firms and Think tanks, which were previously never a part of our problems. Noting this paradox, the author refers to a famous American specialist on the Russian economy. Marshall Goldman points out that although many of these firms specialized in domestic American concerns or arms control, "the vast majority have virtually no experience with Eastern Europe, the Soviet Union, or communism. But they are good at preparing and submitting proposals and rounding up relevant consultants" (Angresano, 1997, p. 180).

Such advisors were not very useful to the CEE economies because they had political rather than consulting-research aims. This is shown by the huge gulf which appeared between those pledges given at the start of the reforms and their real results. Below, we will outline what the real implications are for us of the IMF recommendations. For now, we would like to note that not a single advisor emphasized the necessity of the top priority task of the introduction of the competitive mechanism, without which the other market attributes are not capable of operating. Nobody insisted on the imposition of legislation and lawfulness, and on the following of moral rules and norms; although this framework is obligatory for the market agents.

All advisors – who could more accurately be called agitators – were obsessed with the single dogma: with the demand for market freedoms (liberalization) and the privatization of state property. It was seen as the most important precondition for economic growth and prosperity. But adherents of such an approach have found for themselves a sort of "ideological trap" formed from the major vulnerable point of neoclassical orthodoxy – its neglect of the role of social institutions. The core term of the neoclassical approach is perfect competition. Without market competition other terms and notions are irrelevant. It assumes that the institutions will arise, based on the rational behavior of the individual maximizing his/her utility. Such values are embodied in a very complicated and highly developed system of institutions, rules and norms, including lawfulness, trades unions, business ethics and many other things, which provide the setting without which market competition – of any sort, whether perfect or imperfect – simply can't operate. Such a system of institutions emerged in the West as a result of many centuries of difficult and contradictory history. (See for example: Rosenberg & Birdzell, 1986, McNeill, 1991.)

The majority of CEE countries have a distinctive historical tradition. For example, Russian history, which did not experience the Renaissance or the Reformation, had produced a kind of society where individual values were developed much less than collective, and state regulation restricted almost every aspect of social life. It is obvious that such a society has another institutional setting characterized by other social properties, one in which the individual is not separated from society and so market competition is impeded if not ruled out at all. In such an adverse environment, private property is inevitably distorted and cannot operate smoothly as a market.

In the course of recent reform, the state regulation of the economy was dismantled, but was not replaced with an essentially market system. On the contrary, corruption and crime replaced the previously planned system (according to British criminal counter-intelligence, in Russia today 26,000 organized criminal structures control 40% of the national income). This represents the resurgence of old and traditional norms and values, not all of which take criminal form, of course. The corresponding social setting is based on paternalism and non-economic coercion rather than on an individual's rational behavior.

This was very penetratingly noted by the American professor, Samuel Huntington. He wrote that "as the Russians stopped behaving like Marxists and began behaving like Russians, the gap between Russia and the West widened. The conflict between liberal democracy and Marxism-Leninism was between ideologies which, despite their major differences were both modern and secular and ostensibly shared ultimate goals of freedom, equality and material well-being. A Western democrat could carry on an intellectual debate with a Soviet Marxist. It would be impossible for him to do that with a Russian Orthodox nationalist" (Huntington, 1997, p. 142).

So it would seem that the unpredicted results of neoclassical reforms in Russia is the resurgence of such institutions which are based on private property but which do not allow independent market agents and rule out market competition. The Russian institutionalist Victor Volkonsky shows in his book Institutional Problems of the Russian Reforms (Volkonsky, 1998, in Russian) how such imbalances and disequilibria like the domination of the distribution sphere over production leading to a rate of interest and trade revenue many times higher than the rate of profit in manufacturing; a proportional rise in the raw material sector and the decline of manufacturing; the increase of raw materials and transportation costs in comparison with final production prices; the complete crowding out of private investment from the credit funds market and so on.

Maybe the most important result is the shortage of cash, the greatest in world economic history – the ratio of the money stock to GNP is only 8–10% in present-day Russia, compared to 90–100% in U.S.A and 65% under the Soviets. It has led to an absurd situation when about 70–80% of deals between enterprises of the real sector of the economy are undertaken without real money. Non-payments in the Russian economy have far exceeded the actual money stock. It is a result of the persistent restrictive monetary policy which the Russian government steadily followed under IMF pressure. Can you imagine a market economy without money? Does it make any sense to apply subtle neoclassical models well equipped with sophisticated mathematical apparatus to such barter pre-market economy?

Having no intention to offend western economists, many of whom are well-known as competent and rigorous specialists, we nevertheless have reason to say that official western advisors to the Russian government have greatly damaged the transformation process and the process of the establishment of democracy itself. They share responsibility for it together with the Russian government. At the same time, it would be unfair not to mention other voices among western specialists. James Angresano provides reference to a large range of work representing critical and objective views of the problem. For example, Peter Nolan, cited in the book, gives the following evaluation of the western influence on Russian reforms: "Western social science failed badly in this period. It was deeply flawed. The advice which flowed from orthodoxy contributed substantially to the Soviet disaster" (Angresano, 1997, p. 121). It is difficult to disagree with such a view.

4. CHINESE MIRACLE

In contrast to the CEE countries, China did not follow the assumptions of the neoclassical theory or IMF recommendations "introduced in the cookbook manner" (Angresano, 1997, pp. 28–29), but proceeded to shape its way, tailored on the needs of its population and its own peculiarities. Instead of "shock therapy", it chose the gradual way of transformation, fulfilling it step by step according to the principle of "learning by doing".

The book illustrates the creative character of Chinese strategy. Evaluation of the real situation in the country was derived by the Chinese leadership from a deep knowledge of its specifics and peculiarities. The author notes: "In rejecting the 'cookbook approach' to transformation in favour of treating transformation much as one plays a game of chess, Chinese policy-makers nurtured the development of traditional market-orientated institutions and working rules which suited the Chinese economic, political and social environment. They chose to rely upon common sense and experience rather than upon any abstract theoretical economic model. In the process they developed a transformation perspective which has been relatively less ideological and more measured, experimental, evolutionary and pragmatic than that of the CEE policy makers" (Angresano, 1997, p. 9).

The author does not limit himself just to a comparison of the "Chinese miracle" with the performance of the countries applying the IMF approach, but he analyses the causes of such different outcomes. The Russian reader cannot miss his statement that "it can be forcefully argued that since China faced more difficult initial conditions but has developed a Myrdalian sense far more rapidly than Russia and more than nearly every CEE nation, the problem must lie in the method of analysis and theories relied upon to explain the Russian economy's behavior, and policy prescriptions of the Russian policy makers and their Western advisors holding a neoclassical perspective towards transformation" (Angresano, 1997, pp. 119–120).

In contrast to Russia, which blindly and without any concessions adopted the neoclassical model in its obsolete variant (that is, consistently relying upon the "laissez-faire" principle), China chose the evolutionary way to the market economy, measuring every step like in a game of chess. "Chinese policy makers avoided large-scale reform by experimenting on a small scale at the local level, and only after the experiment was deemed successful would a transformation policy be introduced over a broad geographic area" (Angresano, 1997, p. 9).

Relying on Myrdal and some other authors (Komarek, Amsdem), the pragmatic approach is compared and contrasted with the doctrinaire one, inherent in the "shock therapy", which was not tested anywhere beyond the walls of Cambridge, Massachusetts. The irrelevance of such practice, when the case is reduced to "argument by assertion", the author reveals clearly: "An analyst with a more objective position would have avoided any claims of being unbiased while providing a statement of normative propositions, thereby permitting those who evaluate the analysts' conclusions to test them for logical consistency" (Angresano, 1997, p. 100).

Such a claim to objectivity of analysis is completely in tune with Myrdal's perspective, and the author himself strictly adheres to it. To exclude any bias he insists on the elaboration and application of the indicators, which measured and weighted would help to derive the general index of economic performance. In this way, he considers, we would achieve a reliable tool of objective comparisons, conclusions and recommendations. Such a suggestion seems fairly relevant and meaningful. The only remaining question is whether it will find a broad application.

The objectivity of the author's evaluation of the Chinese results coexists strangely with his underestimation of the theoretical foundations of the same reforms. While appreciating Chinese decision makers for their independence, comprehension and consistence in pursuit of the country's interests, he agrees with Nolan, who sees the reason for the Chinese success in relying on common sense and experience, rather than a high level of economic theory: "rather than regarding the economy as a simple mechanism akin to perfect competition,

Chinese policy makers developed their own transformation programme which featured an eclectic blend of economic philosophies" (Angresano, 1997, p. 25).

These statements seem to me to be not only false, but even to contradict the concept of the work. What does "a high level of economic theory" mean? What are the criteria? We should be very careful in evaluating thought which belongs to another system of values. If we rely only on formulae and graphs, then maybe Chinese theory is not very impressive. But the book emphasizes another perspective – the holistic approach. From this point of view, the matter can be seen in another light.

Nowhere does the book analyze Chinese theoretical reflection per se. That is why I would dare to add a few words about it, recognizing that my own opinion would also be imperfect. To the extent to which it is possible to consider papers by Chinese specialists, published in Russian periodicals, and on the observations of Russian specialists in the Chinese economy, we can note institutions in its formula of the planned market economy which gives a theoretical framework to the Chinese reforms and eloquently explains their distinction to the processes in the CEE countries. Does it really represent "an eclectic blend of economic philosophies"?

In order to decide whether any programme or idea is eclectic or holistic, one has to explore the interrelations of its components. If the latter contradicts the other, then it is eclectic; if all the parts of the system are in tune with one another, then it is holistic.

Let us return to the main directions of the Chinese reforms, as they are set out in the book:

(1) rapid expansion of the private sector driven by entrepreneurial activity
(2) continued government support of the state-owned enterprises
(3) relatively easy access to credit for new enterprises
(4) direct foreign investment on a large scale (Angresano, 1997, p. 25)

Outlining this direction of the reforms, the author virtually partially decoded "the planned marked economy" – the key factor is the interaction of government control and interference and market spontaneity.

We have to note a widespread misunderstanding in the West concerning the notion "plan". It is often identified with rude coercion. In reality it is nothing of the kind; although coercion was inherent in communist societies and interfered in planning as well, it is by no means implied by the notion and the institution itself. The precise meaning of the term is providing proportional economic development. Proportion is a synonym of the more usual economic term, as used in the West, "equilibrium". Of course, the belief that all sorts of proportions can be ensured by the state proved to be false. But the same can be said about the unfettered free market economy, which failed in ruins in the great historical disaster of the Thirties. Western society adopted the experience of national-scale planning under the guise of "Keynesianism" and state regulation of the economy. The structures of imperfect competition embodied in modern manufacturing industries are in many aspects very similar to the organizational structures of the Soviet industrial system. The dual model of the developed industrial economies implies two sectors: manufacturing with costs determined and markedup prices, and raw materials with demanddetermined prices, developed by Kalecki, Eichner, Kaldor, Weintraub and other western economists (Reynolds, 1987). This could also be called a "planned-market economy". Such a system has existed and prospered in the West, contradicting neoclassical values, and since the time of Franklin D. Roosevelt's "New Deal" has added state regulation. Does this make Keynesian and related currents of western thought "an eclectic blend of economic philosophies"?

Incidentally such a blend has a foundation which makes it coherent. In some aspects economic equilibrium is achieved through market forces, in the other – mainly macroeconomic and individual levels of industry – through state and big business regulation. This organic interaction and even unity in some aspects of the planned an market sectors of the economy was elaborated by Russian economic thought in the 1920s when we had a dual economy. It was the first experience of the mixed economy, but these ideas are scarcely known in the West.

It should be added that one of the core principles of the Chinese strategy is an increase in living standards. Of course, to ensure a more equal distribution of income in an increasing market society is no easier than to enhance growth, and it is not achieved every time, but there is a great difference in the two approaches derived from two assumptions. While the essentially neoclassical requirements of the IMF are orientated on the elite strata of society, the Chinese strategy is designed to appeal to the population, to overcome backwardness and mass poverty. Referring to Nolan, the book says: "The absence of human rights is often cited as a strong negative feature of China's reforms. Many characterize the Chinese regime as 'turbid' and 'oppressive'. While virtually no analysts disagree with this assessment, some point out that if a broad definition of such rights is not only the right to vote but the right to an improved standard of living and an environment with a low violent crime rate, then China's 'human rights' record is better than that of Russia and some other CEE nations" (Angresano, 1997, p. 24).

Inspired IMF reforms are designed neither to improve living standards nor to lessen rates of crime. It is obvious that privatization and to the same extent price liberalization were fulfilled in Russia at the expense of the rights of the vast majority of the population. Imposing mass poverty on its citizens, Russian reformers led it to intolerable differentiation thereby undermining the prospects of democracy.

5. OLD FLAWS OF THE NEW EDUCATION

I would particularly like to note the penetrating essay in Angresano's book on the state of economic education in CEE countries. This is a field especially close to me because I am at the head of the Department of Economic Theory and Entrepreneurship in the Russian Academy of Sciences where about 150 choice postgraduates are taught this subject. I can only endorse everything written in the book on this subject. Of course, in comparing the present situation with previous times, one may note that in many important aspects it is better now. Arbitrary prohibitions and obligatory courses were in the past under strict ideological control. Receiving freedom to reform programmes, we have tried to overcome the onesidedness of the previous Marxism–Leninism education system.

But no less obvious is the emergence and quick establishment of another bias: the prevalence of one current strand of Western economic thought, and one which contains a strong ideological component – neoclassical theory. Not only Marxist works deserving attention, but even those of the social-democratic and Keynesian theoreticians, do not occupy an appropriate position in the education curriculum. Keynesianism in the variant of neoclassical synthesis, of course, is included in textbooks. But alternative versions, namely, Post-Keynesian and Radical economics, are not mentioned in these at all. For example, such a famous leftoriented thinker as M. Kalecki is slightly known among specialists. In practical terms, only neoclassical textbooks are available from publishers. In specialist periodicals and published books, the same school of thought prevails. Angresano outlines the flaws of such a system with his usual objectivity and considers them a new version of former Soviet dogmatism:

> Economics education throughout the CEE prior to 1989 focused on techniques of central planning and enterprise management. Stale MarxismLeninism served as the philosophical basis for the economic curriculum, elevated by professors, or "mandarins", of the Soviet Union to the level of state religion in order to indoctrinate students with a "strictly conformist way of thinking" (Myrdal).
>
> After 1989 many western economists received funding to teach undergraduate and graduate students, professors, and policy makers throughout the CEE region. These economists were imbued with the Cold War mentality, caught up in the euphoria of the 198990 political revolutions with a majority adhering to the neoclassical perspective and

the belief in the universal validity of orthodox economic theory. Consequently, CEE authorities, faculties and students were introduced to the neoclassical perspective towards transformation. In all but a few isolated cases, no other perspective which differed substantively from the orthodox view was taught (Angresano, 1997, p. 138).

This is a very accurate assessment. I would only add that a strong and coherent system of funding from Western private and public sources emerged, promoting textbooks and other editions, financing whole educational centers, paying grants to scholars, organizing sessions and workshops and so on. All this overwhelming activity promotes neoclassical traditions without exception and in the absence of any positive national education policy and finance from the Russian Government determines the character of economic education in our country. So far as alternative literature is concerned, not only the social-democratic (Myrdal et al.) but also the wider Keynesian and Post-Keynesian content is practically absent from the libraries and our educational curriculum. In contrast with Soviet times, this is not because of prohibition, but it is not encouraged either and, what is most important, it is not financed. Meanwhile, recognizing this failure of the course based on neoclassical values and partially owing to a distrust of ideas forced upon them, the Russian public – well, at least the intellectuals, but also many politicians and managers as well – have become more and more interested in alternative perspectives. Thus I consider that the work we are discussing would find many eager readers in Russia, but doubt it will find a domestic sponsor for its translation and publication.

6. SOME CRITICAL COMMENTS

Despite a high appreciation of the book in question, and the recommendation of the desirability of its appearance in Russian and other CEE languages, one cannot avoid certain critical comments. That is only natural when we are considering such a complex and contradictory process as transformation with its epochmaking character. For all its achievements, the book is marked by an "outside" view and probably some things are best seen from "inside".

First: Currently the transformation differences between countries have shaped themselves so sharply, it is hardly justifiable to join together all CEE countries in one set with the same or close properties. Poland, the Czech Republic and Hungary are all closer to the West than the others and if they join the European Union then probably they will eventually reach Western standards of living and a Westernized economy. That can hardly be said about a number of other countries, such as Russia, the Ukraine and Belorussia. The second group differs sharply from the first, not only by its current level of development, but, what is even more important, by its ability to adapt to the

Western market culture. Samuel Huntington, in his previously mentioned book, called this group of countries "torn civilizations" – that is, societies containing elements of different civilizations, namely, Western and Orthodox. This point is not new for Russian culture – the core of Russian historiography is the East/ West strands. It can be seen reflected in the great debates between Westerners and Slavophiles in pre-Revolutionary times. One must not forget that Marxism, even Communism itself, arose in the West and, although it opposed Western capitalism, always dreamed of uniting with the socialist West.

In the Soviet period disappointment with communist ideals gradually evolved because we constantly compared our achievements with the West. This was intrinsic to communist ideology and became its "logical trap". Owing to this disappointment, there was an evolutionary shift in the inner orientation of society which gave Gorbachev the potential for reform. But following Yeltsin's reforms, it seems to me, our identification has changed again. The capitalism which has been established post 1989 – not that shown in foreign textbooks and imported ideology but the real one – has been deeply disappointing. One does not speak about the betteroff minority which has gained the position of junior partner of foreign capital, but Russian common citizens who are now likely to be more adverse in their view of the West. If yesterday, the people idealized the West, today they are disappointed with it.

Second: This book, as nearly all other western publications in this field, underestimates the achievements of the CEE nations during the socialist period, thereby impeding understanding of the distinct character and complications of the current changes.

Undoubtedly, the real communism, implemented in Eastern Europe and some other countries, was flawed and committed awful crimes. That is why its deep and comprehensive change became urgent, an historical necessity. Now returning to it is out of the question. But does the condemnation of its sins mean the justification of the flaws of the postcommunist era? If one only views one side of communism then it probably is. But it is worth remembering that great historical phenomena – and communism is of that kind – are usually complex and contradictory. For example, during the history of Western capitalism, violence and human suppression played an important role; colonial wars with millions of victims and slavery; exploitations leading to class struggle and even social upheaval; two world wars and fascism, which was not only produced in Germany but caused by deep crises in Western civilization as well.

Despite all these tragedies, Western society did not reject its values. In the course of economic and social development, step by step it reached social coherence and steadily developed democratic mechanisms. Communism also experienced fundamental evolution since Stalin's time. Without that Gorbachev's Perestroyka would never have happened. Let us ask some pertinent questions: Why do so many countries after several years of liberal reforms experience a socialist renaissance? Why do free elections in Russia, the Ukraine and Belorussia give a majority in parliament to former or current communists? If Communism represented nothing good, then it would be excluded. But the communist past of Kvasnevsky did not prevent his succeeding the liberal Lech Walensa.

Once again, we are not talking about the rehabilitation of communism, but mention this point in order to attract one's attention to the real complexity of the situation. The problem is that in the course of comparing the two systems, not only do the flaws of the current regime reveal themselves clearly but it becomes obvious what a heavy burden it has become for millions of people. It seems that in the West this is often not recognized. One exception is Marshall Goldman, a wellknown specialist on CEE economic affairs, mentioned in the book. Recently he noted in the Moscow newspaper *Novaya Gazeta* that the current Russian regime represents a combination of the worst features of communism and capitalism. This assessment seems to me quite accurate. To prove this statement, we have to provide some evidence.

Despite all its deficiencies, the former planned economy had shown something regarded as impossible before it existed – cohesion of economic growth instead of cyclical development. Does it make sense? Let's refer to the data:

> Since the Bolshevik revolution of 1917, the Soviet Union has transformed itself through an intensive drive for economic modernization, from an undeveloped economy into a modern industrial state with a GNP second only to that of the United States. During that period, the Soviet economy grew more than fivefold. Its industrial structure has changed diametrically (Offer, 1987, p. 1767).

At the start of the five year plans, U.S.S.R. was behind the U.S.A. in the level of per capita income by 56 times, but only by 2 at the end of the Soviet period. But the U.S.A. also did not stand still; it experienced an economic boost. It is true that the Soviet rates of growth decreased in the Seventies and the Eighties, but on the way it had become the world's second superpower. This was not only in the sphere of weapons as many people think. It was a breakthrough in many different areas – education and science, culture and social guarantees. Undisputed areas were free education and medical care available to millions, including cheap facilities in sanatoria and health resorts in the best places. During the summer holidays, a network of free or cheap rest and sporting centers accepted youngsters and children. Municipal transport was practically free, coach services cheap and other transport easily available for the majority of people. There was a shortage of accommodation but apartments were provided free or on low rent. The construction of housing was financed by the Government Habitants of the North and even remote areas enjoyed additional benefits.

Under such conditions and with full employment, it is true wages were lower than in the West but it could not be otherwise. There was no problem in paying wages on time. Payment was guaranteed by law. Under the new capitalism, wages are set arbitrarily and continuously decline in real terms. There are no longer guarantees that wages will be paid at all. The situation over wages illuminates a great misunderstanding of the Russian reforms which is widespread in the West. Attention is always focused on the flaws of the socialist system and its achievements are ignored. So on speaking about Russian problems, many in the West don't take into account what an unfavorable background for market change was created by the real socialist records. This misunderstanding, as it seems to me, was not avoided by Angresano as well, when he wrote that "establishing political democracy does not mean rapid realization of economic prosperity. Nearly all OECD nations facing initial conditions' less favorable than CEE nations, took decades to achieve comprehensive transformations of their economies" (Angresano, 1997, pp. 129–130).

Because of space limits, we will not deviate from the main theme discussing in detail the comparison of "initial conditions" of CEE and OECD countries. And, of course, political democracy doesn't have a direct link with prosperity. I would only like to note that despite their other advantages, OECD countries initially did not possess such important preconditions of "economic prosperity" as the modern industrial potential and highly qualified workforce as CEE countries did. And one should realize that the main criterion of the reforms for the common people is their standards of living and social welfare. At the beginning of the transformation process, as mentioned earlier, Russian authorities gave irresponsible pledges about "rapid prosperity" in several months. The lower social costs and short crisis period of the "shock therapy" in comparison with the gradual approach was one of the main arguments of the Western advisors which received broad publicity. Once again, the complete failure of these obligations had undermined the democratic process in Russia, leading to the Parliament shooting in 1993, the adoption of a rather authoritarian constitution and also to the tragedy of the senseless war in Chechnya. Especially in the light of the Chinese success – and China had much worse "initial conditions", not possessing an industrial base comparable with Russia's – the rapid devastation of the economy and of cultural and social welfare levels achieved under socialism, do not provide very convincing arguments for the market and capitalism.

This underestimation of the market failure in Russia and of its consequences probably occurs owing to the fact that the Western public is not very well informed about the scale of the reform's devastating results. I have to mention it briefly. Already in the whole eight years of reform, decline in production continues and it is much deeper than that which occurred in the USA in the Great Depression. About two-thirds of the capital stock is excessive. The official data of unemployment is unreliable but analysts estimate it in rapidly growing millions. Real wages declined by 23 times. In industry, the wearing out of equipment was 70–80%. Still the "brain drain" continues on a huge scale. Created in Soviet times, sanatoriumcenters are dying because neither state nor citizens have enough money to pay for the services. The mortality rate increased from 10.7 in 1989 to 14.2 in 1996 (on 1000 of the population). The average expectancy of life had fallen from 70 years in 1989 to 66 in 1996. Thus it would not be an exaggeration to say that in the course of the reforms the vital interest of the majority of 90% of the population is sacrificed for the fabulous enrichment of the top 10%.

The future looks even sadder. Investments, which had fallen dramatically by 5–10 times, are stagnant and therefore there is no reason to expect any growth. Governments succeed each other at the expense of growing foreign and domestic borrowing. At the same time, according to Central Bank estimations, the annual leakage of capital amounts to 12–15 billion dollars. In the course of the last crisis this summer the amount soared again. The whole sum amounts to 300 billion dollars (Remnick, 1997, p. 37).

Third: Myrdal's provocative idea of the unity of the socio-cultural and economic (according to Huntington, civilizational) differences determining the course of transformation processes, in my view, is not developed to full logical consistency. I am referring to the elucidation of the different types of capitalism. If capitalism is the same everywhere and the differences referred to above are not strong enough to be generalized in certain typology, then there is no need to criticize the IMF for applying the same model of market economy to all countries. Only if the sociocultural differentials are so significant that they exclude the universal model, should the unified IMF recommendations be criticized.

The usual set of claims that Angresano describes in Poland's example is as follows: "When Poland began to formulate its initial transformation policies, INF support was needed not only to obtain additional financial credit, but to facilitate Polish policy makers' ability to gain credibility from the international community for the implementation of their transformation policies. In response to Poland's request for assistance, the IMF introduced their standard programme with its interrelated components" (Angresano, 1997, p. 7).

The same set of claims was posed before Russia. Let us briefly note their consequences:

(1) Stringent financial and monetary policy. Total money deficit and, hence, an impossibility for firms to pay bills and wages. Half a year, or even much more, delay with wage payments became the norm. The firm has then been forced to move from cash to barter and other nonmonetary systems of payment. That has led to the actual devastation of the unified monetary system, distorted prices, and undermined accounting. Huge opportunities to avoid taxes and make other abuses emerged. All this has led to budget deficits and increasing social discontent.

(2) Achievement of a favorable external balance of payments as well as opening the border to foreign trade, while achieving stability and convertibility of the rouble. This made our exchange rate very unfavorable, opening up loopholes for the flight of capital; induced speculative investments in liquid financial assets imposing upward pressure on interest rates thereby destabilizing the financial market with the constant threat of mass flight; and the influx of foreign commodities depressing the national economy. The result was that the balance of payments became unfavorable.

(3) Establishment of an incomes policy to combat inflation. This is bound to decrease the living standards of the vast majority of the Russian population. Incomes policy as a tool to cope with inflation actually means that real wages decline with the result that the common people have to pay for the incompetence and mistakes of the policy makers. In Russia's present conditions, even the use of this universal method became senseless since the system to a large extent had moved from cash to barter payments. Once price no longer exists, it is no longer a function of supply and demand since inflation measured on such an unusual base cannot be genuine, so its formal decline can mean nothing for the real trends of the economy.

(4) Structural and institutional change. This vague formulation veils the claim for privatization and the establishment of banking and commercial structures. Such claims are suggested instead of real structural changes. These would mean technological modernization, the development of hightech industries and their orientation on the world market, enhancement of foreign exchange revenues which would be the real achievement of the genuine convertibility of the monetary unit.

Such transformation policy, one in appearance and one in essence, was recommended for all the CEE countries. These principles always included the claim of privatization of state property, which, in many cases – and in the former USSR in the overwhelming majority of cases – had adverse results. IMF recommendations are negatively perceived by the author. Chinese success was achieved not only owing to those recommendations but also despite them. His criticism is limited to the Myrdalian positions such as the impossibility of fulfilling the transformation process in the framework of the superficial model of development adopted by one and all. In his time, Myrdal is quoted as saying in an interview with the author: "I've always been on theoretical grounds with this silly idea of models. I mean, for Lord's sake, if you take any American problem, the problem of health, hospitals and all that, you have a quite different history, quite different established situations – you can borrow details, we can borrow certain apparatus or medical instruments, but you can't borrow any model. I've never had that idea" (Angresano, 1997, p. 156). Just the same idea can be applied to the Myrdalian concepts in practice which are inseparable from the phenomena of cultural and social character. Criticism of the IMF by the author on the basis of this reason is relevant on the one hand, but not sufficiently relevant on the other. This must be borne in mind when one asks the question: Why have the CEE countries in general not succeeded in their transformations? But it cannot be the same answer when one asks why China has succeeded.

It is not enough to say that IMF recommendations are false. It is necessary to recognize why one does one thing and not the other. Why does the IMF lead the enhancement of world financial indebtedness and preserve the vicious circle of poverty when it was designed to maintain stability of the world economy?

It seems to me the answer to this last question has to be derived not from the inner logic of the author's own rationale, but also has to be looked at as an example of the dramatic discrepancy between the supreme motives underlying the social and political activities of Myrdal and the real situation he encountered. Once the author asked Myrdal, "Are you now pessimistic or do you feel we've come to where we are today because proper institutional change has not come about?" (Angresano, 1997, p. 160).

For all his theoretical achievements, Myrdal recognized the general failure of attempts to transform the world to a better place. He answered: "Yes, and, of course, the tremendous thing is there is a lack of cooperation – more and more international interrelations and less and less cooperation. Of course, now the world is going to Hell. That is what I am writing about. In the world and in Sweden. And this is a real tragic situation" (Angresano, 1997, p. 160).

Unfortunately the book does not explain this "tragic situation" which Myrdal was so concerned about. Meanwhile, analysis of the outcomes of economic policy promoted by the IMF might enlighten the issue. Of course, the "tragic world situation" can hardly be explained by IMF activity alone. It is likely to be caused by a complex set of factors beyond the framework of the organization and may be even determining its own operations. Nevertheless, its analysis has to show us something important.

At the moment, it would not be an exaggeration to say that the most radical attempt in Russian history to westernize the country – this time on the basis of a coherent neoclassical approach – has failed. The cost has been the actual deindustrialization and the demodernization of the society which, indeed, puts back the prospects for the achievement of democracy and material well-being. Today nobody can predict what tragedy Russia may encounter in the future. But the search for appropriate ways to advance social progress continues. An alternative Myrdalian vision may greatly contribute to the debate. That is why I think that such an honest and comprehensive book as the one written by James Angresano is of particular value as one of the intellectual steps which might eventually help to change things for the better.

EDITOR'S NOTE

Born in 1927, Soltan Dzarasov received the doctorate in economics from Moscow State University in 1969. After working in the Agrarian Institute in Vladikavkaz, Russia, and as an agricultural farm manager, he served as Senior Lecturer and then Professor in the Department of Political Economy in Moscow State University. Since 1972 he has worked at the Academy of the Social Sciences, and since 1986 he has been the Chair of Economics and Entrepreneurship in the Russian Academy of Sciences. He has authored several hundred publications and received several awards, including one on rouble convertibility from a committee chaired by Wassily Leontief. He has been an active participant, along with Andre Sakharov and others, in the Russian democratic movement, including the Social-Democratic Party of the Russian Federation.

REFERENCES

Angresano, J. (1997). *The Political Economy of Gunnar Myrdal: An Institutional Basis for the Transformation Problem*. Williston, VT: Edward Elgar.

Chossudovsky, M. (1997). *The Globalisation of Poverty: Impacts of IMF and World Bank Reforms*. London: Zed Books.

Furman, D. (1998). *Our Strange Revolution*. Moscow: Apycentre; Kharkov: Folko. (In Russian)

Huntington, S. (1997). *The Clash of Civilizations and the Remaking of World Order*. New York: Simon & Schuster.

Krugman, P. (1995). Dutch Tulips and Emerging Markets. *Foreign Affairs*, 74(4) (July-August), 28–44.

McNeill, W. (1990). *The Rise of the West: A History of the Human Community.* Chicago IL: University of Chicago Press.

Offer, G. (1987). Soviet Economic Growth: 1928–1985. *Journal of Economic Literature, 25* (December), 1767–1833.

Remnick, D. (1997). Can Russia Change? *Foreign Affairs, 76*(1) (January/February), 35–49.

Reynolds, P. (1987). *Political Economy: A Synthesis of Kaleckian and Post Keynesian Economics.* Sussex, U.K.: Wheatsheaf Books; New York: St. Martin's Press.

Rosenberg, N., & Birdzell, L. E. (1986). *How the West Grew Rich: The Economic Transformation of the Industrial World.* New York: Basic Books.

Volkonsky, V. (1998). *Institutional Problems of the Russian Reforms.* Moscow: Moscow State University, MSU-Dialogue. (In Russian)

Szostak's ECON-ART: DIVORCING ART FROM SCIENCE IN MODERN ECONOMICS

London: Pluto Press, 1999, xv + 256

This book amounts to a *reductio ad absurdum* of the post-modernist fashion in nihilist economics. Even so, it does contain some commentary worth considering. If the general thesis is chaff, there remain some grains of wheat to be winnowed out.

Contrary to what the infelicitous title might suggest, Szostak's book has very little to do with the history of art contemporaneous with the history of economics, but after some ado over surrealism and cubism identifies modern economics with pure formalism. Eco-art expresses variously the emotions, pleasures and hopes of the artist rather than what the author conceives to be the real world. Reality is the domain of eco-science, which deals with matters of fact not theory. Since economists deal in theoretical terms, they are merely artists engaged in coded rhetoric for their own amusement, reinforcement of their preconceived notions, evasion of pressing real issues, and preserving their privileged academic provisions by excluding the hoi-polloi.

Now on the face of the matter this is an absurd philosophical position. There can be no pure observation without some model to sort out the relevant data. This is the lesson of Immanuel Kant's a priori *synthetic*. Built as it was on Newtonian physics in three dimensional space, the model proved to be

Research in the History of Economic Thought and Methodology, Volume 19A,
pages 189–192.

applicable only to certain restrictive circumstances. That must be true of every model. We don't have a general model of everything, or the reality of "things in themselves." Rather, we must satisfy ourselves with an incomplete scientific explanation that is a functional mapping from a defined domain of data to a range of predicted values. Both the domain and range must necessarily be a subset of the infinite dimensional reality, which lies beyond the ken of finite beings. The demand for absolute, universal truth has no meaning without specifying the range and domain of the explanation. With this functional specification in hand, it is certainly possible to distinguish theories that map from the domain into the range from others that fall outside of it and can be considered false in Popper's sense (Murray Wolfson, 1995).

Szostak does not – cannot – really abandon theory. Rather he tells us that he likes Keynesian theory, and the theories of the classics that he conceives of as dealing in reality rather than theory. He detests the neoclassics, especially those attached to mathematics. The more mathematical the more detestable. Like some nihilist philosopher-critics of economic theory, the most detestable of all to Szostak is general equilibrium theory, that most purely formal of enterprises.

The underlying error is the artificial dichotomy between art and science. Szostak is certainly within his rights to define art so broadly as to include all esthetics, whether or not they fall within the traditional graphic, musical and literary definition of the arts. In that broad sense, all science contains an artistic element. The neat proof, the demonstration that one scientific explanation—say Newtonian physics—is not wrong but a special case of a more general explanation, the beautiful discovery of the DNA double helix are both useful and soul-satisfying. Indeed the more encompassing the model, the more beautiful it is to many observers—although not to Szostak.

The wheat in all this chaff is Szostak's plea for detailed accounts in economics, for tolerance and pluralism and for historical methods. I doubt if he would grant the same tolerance to the econ-artists who he claims have discovered nothing new. Nevertheless, the economics profession needs such critics. Szostak is right that the prevailing methods develop a life of their own that generates a culture and economic incentives that makes advancing heretical views a costly process. At the same time, if he is going to advocate evolutionary methods, he has to generate more than ex cathedra condemnation and citation of authority. He should try to produce scientifically sound analysis that will include these elements. That is not only very difficult to accomplish, but as he says, it is also hard to attract sympathetic attention in the current culture.

Whether that culture generates useful results is a different question. General equilibrium theory, that bugbear of some currently fashionable philosophers, does have important and useful results. Not the least of them is Debreu's demonstration of Adam Smith's conjecture that a competitive economic system would be consistent in structure, efficient in the use of its resources, and optimal in maximizing individual welfare when "each man is the best judge of his own self interest." If reality deviates from Smith's view, then the parsimonious language of mathematics helps to identify the deviation of the data from the model and deal with the problems that arise as a result. At a much more concrete level of abstraction, Leontief's input-output analysis as well as some less theoretically – artistically – satisfying general equilibrium measurements use the same concept in more nearly data driven terms. Clearly there is room for the division of labor in economic analysis.

Apart from the hyperbole, there is nothing really new in this work. The unfortunate aspect of Szostak's book is the lost opportunity to see that both art and science are cultural products. They not only interact with each other, but with historically evolving society, technology and politics. He mentions only the graphic arts, and of these he confines himself to surrealism and cubism. While the author does point out that surrealism was a radical reaction to the First World War, he goes no further than to identify contemporary economics as a plaything of the theorist in order to damn it as surreal.

In fact, a little reflection should have made the author realize that the cultural parallels between the arts and sciences, as well as their philosophical preconceptions, reflect the wider social correlations. After all, while social systems certainly evolve historically, there always must be an underlying element of consistency in any viable society. I have made use of this concept in the teaching of the history of economic thought. Finding – alas – that my students knew very little of this history of anything, I enlisted the aid of my musician daughter, Susan Wolfson. Together we produced *Program Notes to Accompany the History of Economic Thought* (1995). In recordings and visual arts we traced the history of economics from physiocracy to contemporary thought. In that way our students learned intuitively the cultural preconceptions of the great economists as they worked through the original texts – supplemented by only a modicum of lecturing.

Szostak forgets that the arts are a means of communication. If they speak directly to the soul of the scientist or economic agent, then so much the better. To be sure arts and science are not identical even though their underlying sociologies are related. The scientific demonstration of truth or falsity is not the same as the esthetic appearance of beauty or ugliness. Each must stand on its own merits, but the standards of merit reflect time, place and society.

NOTES

Susan and Murray Wolfson. 1995 *Program Notes to Accompany the History of Economic Thought*. California State University, Fullerton, CA.

Murray Wolfson. 1995 *Economia y Filosofia: Lo Que Tenemos es un Fallo en la Comunicacion* (Economics and Philosophy: What We Have Here is a Failure to Communicate), *Cuadernos de Ciencias Economicas y Empresariales, 18*, 27, 97–115. Malaga, Spain).

Murray Wolfson
Professor of Economics
California State University, Fullerton

Barber's THE WORKS OF IRVING FISHER

Assisted by R. W. Dimand and K. Foster. Consulting Editor J. Tobin: London: Pickering & Chatto, 1997. Vol. I: *Introduction. Early Professional Work*; Vol. II, *The Nature of Capital and Income*; Vol. III, *The Rate of Interest*; Vol. IV, *The Purchasing Power of Money*; Vol. V, *Elementary Principles of Economics*; Vol. VI, *Stabilising the Dollar*; Vol. VII, *The Making of Index Numbers*; Vol. VIII, *The Money Illusion*; Vol. IX, *The Theory of Interest*; Vol. X, *Booms and Depressions*; Vol. XI, *100% Money*; Vol. XII, *Theory and Practice of Public Finance*; Vol. XIII, *Crusader for Social Causes*, Vol. XIV, *Economic Policy 1930-1947*. *Index*. Pp. vi, 390; vi, 509; vi, 513; vi, 597; v, 585; vi, 372; vi, 635; vi, 339; v, 612; vi, 350; vi, 312; vi, 537; viii, 301; v, 302.

These superbly edited volumes are a fitting tribute to one of the greatest, some would say *the* greatest, economist that America has produced. Every one of Fisher's five major works and four of his minor works, as well as a dozen or more of his most important papers are reprinted here with explanatory introductions, samples of the original reviews they received, as well as the private reactions they produced among Fisher's friends and colleagues. In addition, many of Fisher's letters to the great and near great are included, which reveal Fisher's relentless efforts to influence policymakers. One is left at the end with a picture of a man whose prolific output must make even the most energetic among us feel positively indolent.

Research in the History of Economic Thought and Methodology, Volume 19A, pages 193–200.

Anyone who writes this much necessarily repeats himself. Even the greatest thinkers only have three or four great ideas in a lifetime. So it was with Fisher. His doctoral dissertation, *Mathematical Investigations in the Theory of Value and Price* (1892) revealed both his lifelong endorsement of general equilibrium theory and his gift for designing mechanical models to represent the automatic adjustment mechanisms of a competitive market. He never had much use for full-blooded cardinal utility theory and quickly adopted Pareto's ordinalism and what was later called revealed preference theory. His first important but relatively short book, *Appreciation and Interest* (1896) invented what is nowadays labelled "The Fisher Relation" – the tendency of the nominal interest rate to rise when prices rise but never to the same extent, so that the real rate of interest typically falls during inflation, apparently because bankers hold unrational expectations – to which he returned again and again in almost every publication for the rest of his life. Next came the first of his major works, *The Nature of Capital and Income* (1906) which developed the fundamental accounting distinction between stocks such as capital and flows such as income, a distinction that is now so much part of the mental furniture of a modern economist's mind that we simply cannot conceive of a time when that distinction was novel. It was this distinction which led him later to advocate a progressive expenditure tax in place of progressive income tax on the grounds that a tax on income taxes savings twice, once when it is earned and then again when it is invested to yield income; since savings are the mainspring of economic growth, it was better, Fisher argued, not to tax them at all (II, 507-508)

A year after *Capital and Income* he published *The Rate of Interest* (1907), in which he showed that the real rate of interest as an intertemporal price could be analysed like any other price as the outcome of the interaction of demand and supply, summed up by "willingness" and "opportunity" lines. His elegant diagrams of how individuals adjust their spending decisions to given interest rates, which then serve to determine the rate of interest in the economy as a whole, managed to capture the highlights of an entire generation's debate of the determination of the real interest rate without invoking Böhm-Bawerk's painful attempt to quantify the roundaboutness of production. He did it so well that it never had to be done again. If one of the meanings of the term "classic" is "definitive", then Fisher's *Theory of Interest* is a classic if there ever was one.

His most famous book, however, is none of these but rather *The Purchasing Power of Money* (1911). All his books display his extraordinary gift of exposition but none better than *The Purchasing Powers of Money*. First, we get the quantity theory of money in words, then in graphic pictures of scales and water tanks, then in mathematical equations, and, finally, the theory is verified by diverse statistics of the relationship between money and prices in different American states in different time periods. Moreover, much of it is set out in something like student notes: there are four forces acting on the velocity of money and each of these is then further delineated; there are five forces acting on the volume of trade and each of these is then discussed; and so on and so on. One gets the impression here, and indeed in all of Fisher's major works constructed along identical pedagogic lines, of a born teacher who never spared himself to drive a message home to his audience. Oddly enough, however, he seems to have been a poor classroom teacher and left few dedicated disciples to carry on the Fisher tradition.

The next masterpiece was *The Making of Index Numbers* (1922), which again was permeated by the sort of manic passion for his subject that we sense in all his books and which almost made the dry topic of index numbers lively and exciting. After laying down a large number of plausible tests that good index numbers ought to possess, he settled in the end for the best or "ideal index number" as a geometric mean between a base-year weighted Laspeyre index and an end-year weighted Paasche index, the bias of the former offsetting the opposite bias of the latter. After 1923, a weekly index number of wholesale prices prepared by his Index Numbers Institute appeared every Monday in *The New York Times* and other papers throughout the United States until 1942 when it was disbanded. In 1995, the U.S. Department of Commerce adopted Fisher's "ideal index number" for calculating the GNP deflator.

The 1920s saw him increasingly concerned about the instability of prices as connected with what he chose to call "the so-called business cycle" – he never did believe that there was any such thing as an endogenous cycle of prices, incomes and output; what there was was a "propagation mechanism", in Frisch's words, that converted exogenous shocks into a cyclical pattern of price and output fluctuations. This led to one of his big ideas, the "compensated dollar", by which prices would be stabilised by ingenious variations in the gold content of the dollar. This gave way in the 1930s to *100% Money* (1935), the "Chicago" plan which would require banks to hold 100% of their check-book deposits on reserve, thus reducing the bank deposit multiplier to unity. This idea of eliminating fractional-reserve commercial banking, thus curing what Hyman Minsky used to call the "financial fragility" of capitalism, became a reform proposal that Fisher advocated with increasing vehemence until the day of his death in 1947.

So the four or five big ideas that became his stock in trade were: (1) the quantity theory of money, (2) the distinction between savings and income, (3) the pivotal role of the imperfectly adjusted nominal interest rate in the generation of business cycles, (4) the endemic instability of prices unless regulated by banking controls, and (5) the debt-deflation theory of depression, which we have not even mentioned yet, which showed that deflation was almost worse than inflation, so that the reflation of prices in the early 1930's became Fisher's principal answer to the Great Depression. We have said nothing about his lifelong battles for alcohol prohibition, vegetarianism, compulsory health insurance, eugenics, calendar reform and international peace. Advocacy of these causes unfortunately gave him the reputation of being a crank, which then raised suspicions about his economic reforms. Moreover, he had gone on record repeatedly to deny the seriousness of the October 1929 crash and to predict imminent recovery every month thereafter for over two years. The fact that he admitted his errors in 1932 (I, 16) did nothing to repair his reputation as a failed prophet. After 1932, he was never taken seriously again either by the American public or by the economics profession. His frantic letters to President Roosevelt throughout the 1930s (XIV, 162ff) are painful to read because it is obvious as one reads them that even Fisher himself no longer believed that his advice would be heeded. By 1946, he even grasped the later Friedman-Schwartz thesis that the stock market crash of 1929 turned into the Great Depression only because the Federal Reserve System adopted a contractionary rather than an expansionary policy in 1930 (I, 24-25), an insight that was years ahead of his time. But by then he was an outsider in monetary policy debates.

One should never read the whole of a man's output at a single sitting or day after day as I have done with these 14 volumes. The proportion of ideas that are endlessly repeated in different permutations and combinations in different books, articles, lectures, newsletters, congressional committee hearings and correspondence with colleagues is extremely high. Nevertheless, the sheer energy and fecundity of his four to five central themes, not to mention the rhetorical ploys that he expertly orchestrated for different audiences, is breath-taking. Even his mechanical inventiveness – the Index Visible Filing System, the Icosahedral World Map and the Portable Stool – make him ten times more interesting than the average economist. Was he the greatest American economist that ever lived (before 1947)? Some might make a claim for Thorstein Veblen or John Bates Clark, but not I.

Let me close with some detailed thoughts arising from reading these pages. Chapter 16 of *The Nature of Capital and Income* contains a suprisingly modern discussion of the risk element on capital valuation (II, 293ff), recommending the standard deviation of dividends as a measure of the riskiness of the returns from an investment (II, 434–438) but without any recognition of fundamental uncertainty in the later Knightian sense. That same prescience is shown in his analysis of the influence of risk on the time preference rates of individuals in

The Rate of Interest (III, 129–132, 237–240). The Fisher Relation is of course discussed in the same book and one is struck by his now standard but then slightly unusual emphasis on the distinction between perfectly anticipated and imperfectly anticipated inflation (III, 108–109, 307–310). Then, too, there is mention of the "reswitching problem" that figured so prominently in the controversies between the two Cambridges in the 1960s, which others have noticed although not until that modern debate was virtually at an end (III, 7, 382). In an appendix to the book, there is a diagram representing an individual's optimal allocation of consumption between two time periods by the tangency of indifference curves and production possibility isoquants with the market price line (1 + r); this 1907 diagram has been called "the trade theorist's sacred diagram" because it appears more frequently in modern international trade textbooks than any other (III, 8, 438–441).

It is one of the tragedies of *The Purchasing Power of Money*, much commented on by Mitchell and Schumpeter (IV, 9, 564), that Fisher placed so much emphasis on the long-run neutrality of money that his extensive discussion of the dynamics of "transition periods" (IV, 95–151; also V, 220–227, III, 2–3) when money is decidedly non-neutral is virtually lost out of sight and was certainly forgotten by subsequent generations. Who now recalls that Fisher conceded the fact that "the strictly proportional effect of prices of an increase in M is only the *normal* or *ultimate* effect after transition periods are over. The proposition that prices vary with money holds true only in comparing two imaginary periods for each of which prices are stationary or are moving alike upward and downward and at the same rate" (IV, 199). It is a tragedy because in the century before him, quantity theorists had gradually come to emphasize the short-run non-neutrality of money rather than its long-run neutrality. Moreover, Fisher spent the rest of his life battling for a compensated dollar, 100% money and the reflation of prices after the 1929 crash, issues which only take on practical significance because the long-run in which money is neutral is only relevant for monetary policy in hyperinflations. Now that the long-run neutrality and even super-neutrality of money is once again standard doctrine in the new classical macroeconomics, we need to underline that old classical and neoclassical emphasis on the short-run interpretation of the quantity theory that Fisher unfortunately minimised.

Fisher designed his *Elementary Principles of Economics* (1912) as an introduction to economics for first-year freshmen. What strikes one almost immediately on reading it is how much of it is macroeconomics. Indeed, out of 26 chapters, only seven treat of the microeconomic determination of prices in product and factor markets. This is not surprising because practically all of it is based on Fisher's three previous books and those were largely focused on the macroeconomic relationships between money, prices and incomes. It is sometimes said that Keynes invented macroeconomics but he did no such thing because the quantity theory of money as Fisher interpreted it was macroeconomics before Keynes.

It is a striking feature of virtually all writings on the business cycle before 1930 that they are not focused on fluctuations of output and employment but rather on fluctuations of prices which are then said to lead to fluctuations in real variables. Thus, counter-cyclical policy in the 1920s was always thought of as consisting of policies designed to stabilise the price level. The titles of two of Fisher's famous articles perfectly exemplify my point: "The Business Cycle Largely a 'Dance of the Dollar' " (1923) and "Our Unstable Dollar and the So-Called Business Cycle" (1925) (VIII, 8–12, 17–40). But the next paper by Fisher, "A Statistical Relation Between Unemployment and Price Changes" (1926) (VIII, 47–56), is unusual not only in emphasizing unemployment but in its uncanny foreshadowing of the Phillips Curve of 1958, complete with a stress on the role of adaptive expectations in generating accelerated inflation. However, Fisher's statistical relation is not really the Phillips Curve because for Fisher unemployment is the dependent variable and inflation is the independent variable, whereas for Phillips the functional relationship is exactly the other way round. The idea of flattening out the business cycle by monetary intervention to stabilise the price level led to one of Fisher's most successful books, *The Money Illusion* (1928), in which he laid out a concrete plan for a "compensated dollar" (VIII, 190–192).

Fisher's *Booms and Depressions* (1932) set out his debt-deflation theory of depressions which ran directly counter to the prevailing view that falling prices in a slump were part of the self-correcting mechanics of a capitalist economy that would eventually induce a recovery. Despite the subsequent label of the "Pigou Effect", it is not at all evident that Pigou himself held this view but it was certainly what Cassell, Schumpeter and of course Mises and Hayek believed (it is a remarkable fact that Fisher never cited either Mises or Hayek and seemed unaware of the so-called Austrian theory of business cycles). The debt-inflation theory was a simple account of why unanticipated changes in the real value of "inside debt" actually aggravate rather than cure a slump and it suffices to show how far Fisher was from a hands-off attitude to the economy, despite an otherwise deeply conservative approach to economic questions (he voted Republican all his life). It is amusing to find him condemning Say's Law – "I do not accept the hoary tradition that general overproduction is impossible and inconceivable" (X, 125, 326) – and approving "an arbitrary program of public works, productive or unproductive or both" (X, 175). But public works

however financed, while unobjectionable as first-aid measures, were condemned as superficial palliatives. The real cure for Fisher was always price
level stability, whether secured by a compensated dollar, or stamped money (X,
344), or *100% Money* (1935), the title of his next book. The trick in adopting
a 100% deposit-reserve ratio was how to move towards it from the then existing
ratios in American banks. Fisher's concrete proposal met with some cogent
criticisms from James Angell of Columbia University (XI, 278–286, 299–306),
which Fisher never successfully answered. He once again stated the debt-
inflation theory but now with increasing vehemence and conviction (XI,
161–170). I am puzzled by the fact that he never provided an effective
exposition of the Bank Credit Multiplier in a fractional-reserve banking system,
contrasting the powerlessness of individual banks with the potency of the entire
banking system to multipy credit (a single footnote citing the work of Chester
Phillips, XI, 809n, is the only hint he ever gave that he understood the point
perfectly well).

1927 saw the publication of one of Fisher's most famous papers emanating
from his interest in public finance: "A Statistical Method of Measuring
'Marginal Utility' and Testing the Justice of a Progressive Income Tax" (XII,
11–49). It contained an elegant thought-experiment to operationalise cardinal
utility scales but if and only if utility functions are additive, that is, if all goods
are independent of one another. Having demonstrated that such utility functions
display diminishing marginal utility, Fisher went on almost casually to add
interpersonal comparability of utility to arrive at the standard Marshall-Pigou
conclusion that an expenditure on income tax ought to be progressive. It is
good to be reminded of the "old" welfare economics in the days before
Robbins, Hicks and Kaldor.

Vol. XIV in this collection brings together various writings on the depression
of the 1930s and the postwar adoption of the Full Employment Act, including
a large number of letters to President Roosevelt in a hectoring tone that speaks
well of Roosevelt's patience with self-appointed advisers. It is fascinating to
notice that Fisher confirmed the Kuhnian rule that older scientists rarely accept
a new scientific paradigm: in an exchange of views with Richard Lester in 1936
on wage cutting, Fisher showed no comprehension of Keynesian economics
and clearly never read *The General Theory* then or later (XIV, 157–161,
217–218, 238–241).

The editorial work on those 14 volumes is a model of its kind: the
introductions say just enough and not too much to motivate the items included,
and the choice of letters to reprint from various archives serves to illuminate
Fisher's writings and to bring home to the reader how closely he was in touch
with other economists and politicians around the world. William Barber, Robert

Dimand, Kevin Foster and James Tobin are to be congratulated on a thankless task skilfully executed. These 14 volumes should be read in conjunction with R. L. Allen's *Irving Fisher. A Biography* (1933), a work as long on Fisher's outward life as it is short on his inner thoughts, and H.-E. Loef and H. G. Monissen, eds., *The Economics of Irving Fisher* (1999), a collection of valuable essays on every one of Fisher's writings.

Mark Blaug
University of Amsterdam
The Netherlands.

Blaug's THIRD EDITION OF
WHO'S WHO IN ECONOMICS

Warren J. Samuels

A review of Mark Blaug, *Who's Who in Economics*. **Third Edition.**
Northampton, MA: Edward Elgar, 1999. Pp. xx, 1237.
ISBN 1-85898-886-1. $350.00.

Mark Blaug's *Who's Who in Economics*, now in its third edition, is one of the
approximately two dozen reference works at my disposal in my study. It is one
of the most important for my varied purposes, along with *The New Palgrave*,
the *AEA Membership Directory*, and the various editions of the *Index of
Economic Articles*, plus two other works by Blaug (1985, 1997). Two types of
entries are included: those on prominent deceased economists, the entries
prepared by Blaug, and those on prominent contemporary economists, selected
on the basis of citation counts, the entries prepared by the individuals
themselves.

The first edition, published in 1983, included 700 living and 400 deceased
economists. The second edition (1986) was expanded to include 1000 living
and 400 deceased economists. This edition includes 1400 living economists;
the number of deceased has increased, through mortality, to about 500. The
period covered is 1700–1996 ; thus, for example, no entry for William Petty
(1623–1687) is included, though there is one for John Locke (1632–1704).

Research in the History of Economic Thought and Methodology, Volume 19A,
pages 201–204.
2001 by Elsevier Science B.V.
ISBN: 0-7623-0703-X

Blaug compares these numbers with an estimated 36,000 and perhaps as many as 40–45,000, economists in the world, not all of whom, of course, publish either at all or regularly in the top 200 economics journals (pp. vii–viii). Interestingly, some 80 frequently cited economists did not return the form on which their entry would be based; the group includes several well-known economists (Appendix 4). The first edition (1983) ran to 449 pages; the second (1986), 958 pages; this edition, 1259 pages.

Blaug refers to "The sheer slog of producing a book like this" and to "the amazing scope and spread of interests among practicing economists" (p. vii). One has to concur on both points.

The strength of the book is that it provides a great amount of information on a large number of professionally important and, indeed, interesting economists. Particularly invaluable are individuals' own identifications and/or assessments of their principal contributions and Blaug's summations of the careers of earlier deceased economists. Important personal information is also given.

That the collection is useful goes without saying; it is a valuable resource. Of interest, too, is the question of exclusion. It could go without saying that Blaug could not be expected to include every reasonably possible candidate for inclusion, either those deceased or those currently active. The number of deceased grows with time, as does the ranks of widely cited younger economists. I write "could go without saying" because a case could be made for a much longer list of both groups. Still, Blaug would have to draw his lines somewhere. To his credit, the numbers of included economists has increased. As it stands – I say this as fact and not in criticism – the two groups are elitist; lesser – lesser remembered or valued and lesser cited – economists are not included and, therefore, one has to seek information on "who was/is not quite who" elsewhere. That is to say, the collection could be even more useful but one cannot demand either the impossible or the overly costly, especially in terms of Blaug's time – and at $350.00 the volume is already priced beyond the personal-copy reach of most, asymptotic to all, practising economists. (This is not a matter strictly of transcendent market economics but of pricing policy – one factor in the formation of actual markets. Still, I do not know the elasticity of demand relevant to a much cheaper paperback edition.)

In a review of the book issued on the Economic History Net (ehreview@eh.net) on February 11, 2000, Robert Whaples notes both the strength and weakness of journal citations as a selection device, particularly in regard to those whose contributions are to be found elsewhere, as well as serious omissions from among the deceased. Whaples calls attention to the inclusion of only three of the first eighteen presidents of the Economic History Association, that many earlier economic historians are in danger of being forgotten, and that many recent presidents are missing as well, but notes that about one-third of all EHA presidents are included and that entries for many other economic historians are also included (four failed to reply). Whaples concludes that a comparable volume for economic historians would be valuable. Of course, comparable problems of exclusion would also arise in such a venture.

Not everyone who I have looked up in the earlier editions was included. Indeed, the first two names I looked up in the third edition – Walton H. Hamilton (1881–1958) and Charles H. Hull (1864–1936) – are omitted. Hamilton was a major, albeit neglected, early figure in Institutional economics and in the early history of law and economics; Hull wrote on a number of topics, publishing a well-known collection of the economic writings of William Petty (Hull, 1889), other items in the history of economics (e.g. Hull, 1886), and a piece on the service of statistics to history (Hull, 1914) (his papers are at Cornell University).

I compared the coverage of several randomly chosen specialized works with that in Blaug's third edition. From a survey of economic thought in the Netherlands (van Daal & Heertje, 1992), the major figures – Tjalling Charles Koopmans, Nicholaas Gerard Pierson and Jan Tinbergen – but not lesser names are included in Blaug's collection. Of twelve twentieth-century Italian economists featured in another survey (Meacci, 998), only two are not found here. A casual examination of a Companion to Austrian Economics (Boettke, 1994) and a history of game theory (Weintraub, 1992) indicates that most but not all relevant names have entries in Blaug. (Two of the foregoing four volumes were published by Blaug's publisher.)

Blaug has an entry for Herbert Simon but not Trygve Haavelmo, John Nash or Myron S. Scholes, all Nobel Prize recipients. Perhaps Nash equilibrium is so much a conventional term of art, it neither needs nor receives citation.

A test comparable to Whaples' is to look at Appendix 1, an index by principal fields of interest, and see how representative are the lists for specialized fields measured against one's knowledge thereof. Doing so for two fields, history of economic thought and methodology (B0 through B4), I reached two conclusions: first and principally, that numerous leading names are not present, and second, a couple of names are surprisingly present and a couple, otherwise present, are not listed. The former is due to the citation procedure; the latter, to self-identification. As with Whaples, the result is that one can learn a good bit about a field from the collection but by no means all that one might want to, or should, know. (For the record, to declare my interest, I have been included in all three editions.)

Even exclusion can be useful. One of my uses of the second edition was to help in confirming which continental economists working in the late nineteenth and early twentieth centuries, in part by virtue of their *exclusion* from Blaug's collection, could be deemed neglected (Samuels, 1998).

The conclusions must be that this is a *very* and (as between editions) *increasingly* useful volume; that, like any comparable venture, omissions are inevitable – some more embarrassing than others; that the omissions are selection-criteria specific; and that some uses are more valuable than others. Indeed, embarrassment may not be the correct word; having adopted the selection procedure, one is prepared to accept the results, however disconcerting some may be. Mark Blaug, perhaps the most prolific scholar, certainly one of the foremost scholars, in the fields covered by this annual, is to be congratulated not only for keeping up his good work but for making it better.

REFERENCES

Blaug, M. (1985). *Great Economists Since Keynes*. Totowa, NJ: A Barnes & Noble.

Blaug, M. (1997). *Great Economists Before Keynes*. Northampton, MA: Edward Elgar.

Boettke, P. J., (Ed.) (1994). *The Elgar Companion to Austrian Economics*. Brookfield, Vt: Edward Elgar.

Hull, C. H. (1896). *Graunt or Petty? The Authorship of the Observations Upon the Bills of Morality*. Boston: Ginn & Co.

Hull, C. H. (1899). *The Economic Writings of Sir William Petty together with the Observations Upon the bills of Mortality, More Probably by Captain John Graunt*. 2 vols. Cambridge: Cambridge University Press. Reprinted in 1997 by Routledge/Thoemmes Press.

Hull, C. H. (1914). The Service of Statistics to History. *Quarterly Publications of the American Statistical Society*, vol. *14* (March), 30–39.

Meacci, F. (1998). *Italian Economists of the 20th Century*. Northampton, MA: Edward Elgar.

Samuels, W. J. (1998). *European Economists of the Early 20th Century*. Vol. 1. Northampton, MA: Edward Elgar.

Van Daal, J., & Heertje, A. (Eds) (1992). *Economic Thought in the Netherlands, 1650–1950*. Brookfield, VT: Avebury.

Weintraub, E. R. (Ed.) (1992). *Toward a History of Game Theory*. Durham, NC: Duke University Press.

Pasinetti and Schefold's THE IMPACT OF KEYNES ON ECONOMICS IN THE 20TH CENTURY

Melvin W. Reder

Edited by Luigi L. Pasinetti and Bertram Schefold. Cheltenham, U.K. and Northampton, MA: Elgar. 1999. Pp. xvii, 239. $96.00. ISBN 1-85898-861-6.

This volume consists of 13 papers presented to the first Annual Conference of the European Society for the History of Economic Thought held in early 1997. These 13 were selected from more than twenty-five papers that had been presented and discussed. The book is divided into three parts: I (Theory) consists of three essays on Keynes's thought, mainly The General Theory (GT) by Pasinetti, Leijonhufvud and Skidelsky; II (Keynesianism in European Countries ...) "aims to present an overall picture of Keynesian-type or Keynesian-like policies and theories that were proposed in the major European countries – not always and not necessarily stimulated by Keynes"; III (Institutional Dicussions of Keynesian Policies) consists of a series of papers showing "how Keynesian ideas – mainly about policy – have either penetrated, or have emerged from ... debates and controversies ... going on inside governmental, intergovernmental or international committees ...". After a

Research in the History of Economic Thought and Methodology, Volume 19A, pages 205–217.
Copyright © 2001 by Elsevier Science B.V.
All rights of reproduction in any form reserved.
ISBN: 0-7623-0703-X

general comment I shall remark upon particular features of II and III, leaving I to the last.

To judge the impact of Keynes upon economics it is necessary to remember that the peak period of his activity, 1930–1945, roughly coincided with two other major developments: (1) a big step toward formalization and quantification of economic reasoning and (2) the spead of English as the common language of the economics profession. The discussion in all of the papers of Part II serves to remind us of how much less technical a subject economics, especially the macro part, was in the 1930s than it is now (at the beginning of the 21st century). The far greater salience of mathematics in the exposition of theoretical argument is expressive rather than constitutive of the transformation that has occurred. The substance of the transformation lies in the focus of contemporary professional discussion upon the characteristics of explicitly stated models that make, at most, only incidental reference to events in specific times and places. This focus upon the abstract is symbiotic with use of a mathematical format, though neither requires the other.

This is not to suggest that the earlier type of discussion, specific with regard to time and place, has disappeared. Indeed, appearing in non-specialist journals, publications of governments and international organizations, financial newspapers and semi-popular books, its quantity almost certainly has risen. And a substantial part of it (including contributions by technically proficient economists on holiday) is of high quality as judged by contemporary standards. However, this work is usually considered as popularization and often is accompanied by technical appendices where the argument is set forth in a professional style (usually involving some mathematics).

It is the comparative lack of such appendices and/or the felt need of argument at the technical level (revealed by footnote references to appropriate sources) that distinguishes the writing of Keynes's contemporaries, and of the pre-GT Keynes, from that of contemporary economists. GT itself represents a transition from the older style to the newer: much of it is focused on the specifics of the U.K. in the 1930s and could be understood by the informed public. But other parts – those serving as jumping off points for the debates of the late 30s – are of a different genre making explanations of observed phenomena dependent upon the properties of formal models.

Whether it is essential to employ mathematics to establish the logical consistency of a model remains moot, but it soon became apparent that communication on the relevant issues was much easier, and arguments more persuasive, when presented in mathematical terms. Here, the Keynesian Revolution became confluent with the rise of econometrics. Formed in 1931, The Econometric Society aimed at both the integration of economic theory and statistics under the aegis of mathematics, and the internationalization of economic science. Quickly it attracted a group of energetic young economists interested both in mathematics and in the kind of macro problems which GT addressed. Immediately upon its publication, GT became a focal point of interest for this group who sought to interpret the work so as to make it mathematically tractable. Increasingly their discussions set the tone of debate about the professional status of GT.

Although mathematically expressed general equilibrium models had existed since Walras, they had been considered a curiosum and kept apart from the mainstream of political economy that addressed questions of policy. Moreover, such models did not envisage the possibility of an economy with unemployed resources. By thrusting this possibility into the forefront of discussion, GT compelled an augmentation of the Walrasian framework to permit its consideration. Proper formulation of this augmentation has been an important preoccupation of economic theorists for over half a century and has been conducted almost entirely as a mathematical exercise.

Partly as the result of the Econometric Society's international orientation, and (perhaps) even more because of the flight of European intellectuals to England and the United States during the 1930s and 40s, and the superior opportunities for employment and training in the U.S. (at least through the third quarter of the last century), mathematically laced English became the lingua franca of technical economics. After 1935, regardless of nationality, among aspiring economists the best and the brightest published in this language and visited and often remained to teach in the U.S.. The effect was to create an international profession and literature that had hardly existed prior to the 1930s. To attribute this development entirely to the influence of GT would be an exaggeration, but the symbiosis of these coterminous developments cannot be denied and the impact of GT upon economics is felt as part of their joint product.

Yet a further aspect of the confluence of Keynesianism and econometrics was the development of National Accounting. Though the statistical work (by Colin Clark and Simon Kuznets) that led to the creation of these basic statistics was quite independent of the theoretical ferment that brewed GT, the conjugate relation of the two is obvious. GT made use of Colin Clark's early calculations and American Keynesians were quick to clothe their theoretical framework with Simon Kuznets's numbers, and to lead the cry for more of them. It was soon perceived that proper use of these numbers in a multiequation model would require development of new statistical techniques which had been a goal of the Econometric Society from its inception. This led to the development of techniques for estimating multiequation models of the economy by economists associated with the Cowles Commission. These models were implemented empirically by economists whose Keynesian inspiration (e.g. Lawrence Klein) is indisputable.

That Keynes was not very sympathetic to the development of mathematical economics, and even less favorably disposed to econometrics, is well known but irrelevant. The generation of economists throughout the world who were inspired by The General Theory were also excited by the prospect of making economics either a mathematical or a quantitative science, or (very often) both. Thus, until well into the 1960s, econometric models of national economies and the underlying economic theory had an unmistakable Keynesian flavor. Since the 1960s, both in economic theory and in econometric applications, there has been a neoclassical resurgence which is non or even antiKeynesian in spirit. However, this work is no less mathematical or quantitative than that done by contemporary economists in the opposing new or neo Keynesian camp(s).

Thus the split among economists about the proper role of mathematics and quantitative methods no longer parallels a division between Keynesians and antiKeynesians. But this should not obscure the importance of the parallel of these divisions during the early history of GT and the complication that it presents to anyone seeking to appraise the impact of Keynes upon economics. Although this confluence of developments hardly qualifies as news, no recognition of it is to be found anywhere in this book.

Now let us discuss Part II. The editors state in the Preface (p. xii) that they have decided to omit discussion of the Keynesian connections with Swedish economic theory and Michal Kalecki, both because of the vast literature on these connections and in order "to give more space to other, less known, connections – in France, Germany, Italy and Spain". Avoidance of redundancy aside, the consequences of this editorial decision are unfortunate. With a few exceptions to be noted, the papers of this section are superficial. In good part, this is because the authors try to cover too much: one learns of the views of economists (obscure to English readers) on various arguments of GT from brief quotations with scant attempt – probably because of spatial constraint – to present the supporting argument. From what this reviewer can see, English-only readers have not missed much, but this judgment may be unfair.

The French authors divided the Keynesian era into two (overlapping) periods, 1919–1931 and 1929–1969. The first period is discussed by Alain Alcouffe and the second by Richard Arena and Christian Schmidt. Alcouffe's report is focused on two authors, Charles Rist and Jacques Rueff. Contrary to the attitude of the French public and the official position of their government, Rist took a favorable view of The Economic Consequences of the Peace and the associated plea to reduce the level of German reparation payments. However, he disliked both The Treatise and The General Theory because of his belief in Say's Law and a general adherence to a Walrasian perspective. From Alcouffe's presentation , it would not appear that, even had he written in English, Rist would have contributed much to the debate about the merits of GT.

Jacques Rueff more or less shared Rist's affinity for Say's Law and a belief that price flexibility would always cure unemployment or indeed any problem of (seeming) market failure. Alcouffe presents in some detail Rueff's statistical argument (including his data) that during the 1920s, unemployment in the U.K. was caused by high real wages with the dole as a contributing factor. As presented, the argument amounts to treating a simple bivariate intertemporal relation between unemployment and real wage rates as a causal relation without any consideration of the roles of aggregate demand or other variables: enough said.

Arena and Schmidt discuss a much wider group of economists than Alcouffe, but the views of none of them are described in detail. I would have preferred to learn more of Maurice Allais's ideas on GT, but Arena and Schmidt distributed their attention in a quite egalitarian fashion. Although much too lightly, Arena and Schmidt do make a few points of general interest:

(1) in remarking on Emile Borel's comments on Keynes's Theory of Probability, and their connection with his (Borel's) later work on game theory, they conclude as follows (p. 82): "Unfortunately, Borel was not read by the French academic economists and his audience was limited to a small group of mathematicans and economic engineers, who, later on, were involved in economic analysis". This failure of communication is symptomatic of an institutional idiosyncracy of the French system of higher education that affected far more than the appreciation of Keynes. It is regrettable that Arena and Schmidt did not go into this matter more thoroughly than they did.

(2) In explaining the initially unfavorable reception of GT among French economists, Arena and Schmidt remark that "The hostility which prevailed within the French academic establishment as regards Keynes, and derived from his *Consequences of the Peace,* was still present when the *General Theory* was published in the first English version. (p. 84). They distinguish between "Two different types of reactions, . . . Traditional economists criticized Keynes, sometimes heavily, but their theoretical background was often too weak to give birth to profound comments. . . . Neo-marginalists were less numerous but they had at their disposal a complete and consistent alternative theoretical framework. This is why their criticisms are far deeper". (p. 85) (The "neo-marginalists" include Allais and Divisia.) Accepting their perceptions of relative depth, one would like to know more of the bases for this appraisal. The implications of the weak theoretical background of numerous French

economists which Arena and Schmidt report (and which I find credible) for the organization of the French academic system, and the reforms that may have occurred subsequently, would be well worth an extended account. Perhaps because they chose to limit their purview to the period before 1970, the authors fail to mention the distinctly neoKeynesian writing of some French mathematical economists, notably Edouard Malinvaud and Jean-Pascal Benassy, appearing in the late 1970s and early 1980s. It would be interesting to hear of the reception of this work in France, and the extent to which that reception is still impeded by the communication difficulty noted under (1).

(3) Like the other contributions of Part II, this paper suffers from the unfortunate identification of the history of economic thought with the thinking of professors of economics. This has the effect of obscuring much of the economic thinking that bears upon the making of history. For example: a very important aspect of French economic thinking in the past half century has been associated with the economic unification of Europe. Yet, following the dominant trend in dogmengeschicte, Arena and Schmidt completely ignore the bearing of Keynesian ideas upon the promise and problems of a common currency, an area where the influence of French ideas about economics has been at its greatest.

A similar complaint can be made about the brief paper of Daniela Parisi and Claudia Rotondi on Italian economic thinking which is focused on the work of Francesco Vito and Paolo Seraceno. The concern of these economists with Keynes is not made clear, and Parisi and Rotondi seem more concerned with drawing attention to their existence than to discussing Keynes's impact upon their thinking. As presented in this paper the primary concern of both writers was the promotion of economic development in Italy, mainly through the increase of investment. Seraceno was heavily involved in SVIMEZ (Southern Italy Development Fund), a quasipublic organization concerned with studying the problems of industrializing Southern Italy and finding capital to promote that objective. To appreciate the significance of Vito and Seraceno it would be necessary to know more about the functioning of this organization. Regrettably this paper does not offer the needed instruction, but for this the authors may be forgiven as they take very little of the reader's time: their essay absorbs only slightly more than five pages.

The other essay on Italian thought by Piero Bini and Antonio Magliulo is focused sharply on the influence of Keynes. Neither the Treatise nor GT was favorably received by the majority of Italian economists for a variety of reasons, of which the most important was the belief that the principal barrier to full employment in Italy was deficiency of physical capital rather than lack of effective demand: hence, to increase employment, public policy should aim at a higher saving ratio to facilitate investment rather than a lower one to stimulate effective demand. This led Italian economists, "liberals" and interventionists alike, to argue that GT was a special case, perhaps relevant to "Anglo-Saxon" countries, but irrelevant to Italy. According to Bini and Magliulo this attitude which prevailed in the 1930s continues to the present, even among those otherwise sympathetic to Keynesian views.

As presented, Italy is an interesting case where the specifics of a nation's economic position may have given rise to an idiosyncratic preanalytic vision among a nation's economists. But before accepting this view, I would like to see more of the relevant details. It would help to relate the concern with undersaving to the chronic problem of the Italian fiscal deficit and to the expressed views of Italian economists on this subject. The antipathy of Pasinetti (expressed in a quotation, p. 131) to invocation of "Keynes" to justify these deficits would suggest that Bini and Magliulo may be on the right track, but a more detailed argument is needed.

Finally, Bini and Magliulo do not mention the English (i.e. Cambridge) connection of Italian economics. Starting with Piero Sraffa, and continuing to the present day, there has been a stream of Italian students to Cambridge, many of whom have returned to Italy and taken academic positions. Surely their views have played a role in determining the Italian reaction to Keynes. The interaction of these "Cambridge-Italians" with the Italian academy would make a most interesting story: I, for one, would love to read it.

As indicated in its title, Salvador Almenar's essay "Keynes's economic ideas in Spain before the *General Theory* . . . " deals with the period before the Franco-led rebellion. The principal figure in the paper is German Bernacer who made (and has had made by others on his behalf) claims of having anticipated GT. As presented here these claims would seem to be of limited validity, a judgment in which Almenar appears to concur (p. 112). There is also a brief description of an ongoing debate in the 1920s and early 30s of the merits of fiscal activism in which the name of Keynes is sometimes cited: the names of some of the participants in this debate are given but their arguments are not described in sufficient detail as to justify comment.

Harald Hagemann's essay on ". . . Keynes and economic 'activists' in pre-Hitler Germany" is not a discussion of the impact of Keynes on German economic thought, but a (good) brief discussion of the ideas of a particular group of German economists on the relation of wages and employment during 1929–32. These economists (Gerhard Colm, Emil Lederer, Alfred Kahler, Adolf Lowe, Jacob Marschak , Hans Neisser) were on the faculties of Heidelberg and Kiel and, politically, were Social Democrats. All of them emigrated from Germany in the spring of 1933 and almost all of them were at one time or another associated with the New School for Social Research in New York.

Their general position was that reducing money wages was an ineffective way of combatting unemployment. In varying degree, their arguments stressed the importance of maintaining effective demand and the adverse effects that general wage cuts exert on this variable. Their arguments were similar in thrust to Keynes's on the same topic, though probably formulated independently. However, though still valid on their own assumptions, they are now obsolete. In addition some members of the group (especially Hans Neisser), were also concerned with the interrelation among real wages, capital accumulation, technical progress and the level of unemployment. While the discussions on this interrelation may have been quite advanced for their time, they are simply inferior to what has been available for at least the last 50 years, as (I suspect) most of those involved would have been among the first to admit.

Part III is a mixed bag. Dorothee Rivaud-Danset's essay on the minutes of evidence given to the (British) Macmillan Committee of 1929–31 (Keynes was a leading member) is a gem of thought history. The Committee was charged with investigating the causes of the ongoing depression and the proposal of remedies. Among the remedies proposed, wisely, the author has focused her attention upon one; "rationalization" of industry.

Loosely, rationalization was an amalgam of government sponsored schemes for cartelization of various industries, with the aim of concentrating productive activity in the most efficient plants and closing the rest. The anticipated result was an increase in profits for survivors, shared somehow with owners of abandoned facilities. Hopefully, the resulting unemployment would be offset by an increase in investment resulting from the anticipated increase in profits. Such schemes always involved participation of banks both as sources of new finance and, through their creditor position, as instruments for exerting pressure on reluctant participants. As with the NIRA in the U.S., this was seriously proposed as an alternative, or at least a supplement, to a policy of curing unemployment by increasing effective demand through monetary-fiscal action.

This essay presents the views of British economists (Keynes, Hawtrey, Pigou, Robertson et al.), bankers and industry leaders on (what we would now call) macroeconomic policy circa 1930. Although there are definite traces of the "cure unemployment by cutting wages" view, the ideas expressed were more diverse, complex and confused than a reading of GT might suggest. One leaves this paper with the feeling that there is much more to be learned about the economic thought of the 1930s than is conveyed by a recounting of the progress of economic analysis.

Robert Dimand ("The Beveridge retort: . . .") is an interesting account of the evolution of Beveridge's thinking on macroeconomics. Deeply hostile to the application of economic theory of any kind to the solution of economic problems, prior to GT Beveridge had been considered the leading British authority on the causes and cure of unemployment. In his view the causes of unemployment could be fathomed only through empirical study and therefore, because its argument was purely speculative, GT should be disregarded. In particular, he dismissed the idea that deficiency of effective demand was relevant to unemployment. As a result Beveridge's initial reaction to GT was hostile in tone and uncomprehending of substance.

As Dimand tells it, this attitude changed gradually during World War II, culminating in Beveridge's (1945) acceptance of the thesis that maintenance of full employment required utilization of monetary-fiscal policy to maintain effective demand. Indeed, he went further than Keynes to argue that in order to insure continuous full employment it might be necessary at times for government expenditure to "overshoot" the target. In such an eventuality he proposed price controls and rationing to avoid inflation.

This essay recounts a very interesting episode in the history of economic thought. But it should be noted that Beveridge's change of mind – and heart – was associated with a similar change of general attitude within the British establishment. Some discussion of the connection of these correlated changes would add to our understanding of Keynes's impact on economic thinking.

Anthony Endres and Grant Fleming ("The ILO and the League of Nations: . . .) give a lucid account of the economic research done by these two related organizations during the 1920s. Within the small compass of 15 pages, they manage to convey a good sense of the macroeconomic concerns that animated this research and of the general policy line (stabilization of the price level through monetary policy with maintenance of full employment as a secondary byproduct) advocated. However, they do not investigate the relationship between the characteristics of the sponsoring organizations. This is unfortunate: there is a clear need for study of the effect of the structure of contemporary international organizations on the research that they do and/or sponsor, and the overall effect that this has had upon economic thought and policy. It would also be valuable to have a study of the role of professional economists within these organizations. Comparison of such studies with an appropriate extension of the Endres-Fleming paper would be very useful.

Gerard Dumenil and Dominique Levy ("Pre-Keynesian themes at Brookings") make an apt comparison of the writing of Harold L. Moulton and Keynes. Moulton (leading author of a four volume research series published by Brookings in the mid-1930s and predating publication of The General Theory) was a strong proponent of an underconsumptionist view of the Great Depression. Without attempting even rudimentary econometrics, the Brookings volumes emphasized quantitative description alloyed with (very) loose theoretical argumentation and had much of the flavor of what was later called "hydraulic Keynesianism". Dumenil and Levy present a good summary of their argument which highlighted the contention that too much of the benefits of technical progress was retained as profits – and saved – with the result that the income of wage earners and others was insufficient to support a level of consumption adequate to induce sufficient investment to absorb the savings generated.

In other hands, often such a preanalytic vision has served as a rationale for progressive taxation to redistribute income in order to eliminate the deficiency of consumption. But as the authors point out, Moulton did not react in this way. Rather he called for more competition in product markets to redistribute income (and reduce the savings rate) through lower product prices. However, this did not lead, as well it might have, to a demand for vigorous antitrust action. As Dumenil and Levy comment (p. 194), "Moulton's views concerning competition are not easy to disentangle". Apparently, he coupled a drastic diagnosis of the situation of the American economy in the early 1930s with the hope that mild remedies would suffice for a cure and did not face up to what his diagnosis might imply. In this he was not unlike Keynes, as Skidelsky sees him (see below).

Moulton seemed to consider the situation of the U.S. in the 1930s as a rare event, possibly justifying temporary deficits but definitely not providing a rationale for a secular rise in the National Debt: this is duly noted in the paper. However, the authors neglect the opportunity to draw attention to the sharp contrast of this view with the position of Keynes in "How to Pay for the War" (1940). This omission is shared with almost all the contributions to this volume. Aside from a brief incidental reference in Dimand's piece on Beveridge, "How to Pay for the War" is not mentioned anywhere in the book. Yet it is a fundamental link between the thinking of Keynes and postwar Keynesianism. While GT said nothing directly about National Debt, in How to Pay Keynes proposed that, varying with their wealth, individuals should be compelled to purchase government bonds and allowed to sell them only at the option of the Treasury, whose decisions were to be governed by its policy objectives, especially those concerning the price level and the level of employment.

These proposals, and the attempts to implement them and/or to resist their implementation, set the tone of fiscal policy discussion both in the U.K. and the U.S. through the third quarter of the last century and well into the fourth. It is unfortunate that nowhere in the book is this aspect of Keynes's influence discussed because the discussion of fiscal policy – by economists and others – is an important episode in the history of economic thought in the 20th century. Herbert Stein's "The Fiscal Revolution in America" is an excellent discussion of this episode through 1967: a sequel dealing with the counterrevolution that began almost immediately upon the book's publication is badly needed. Such books could serve as the basis for integrating dogmengeschicte with economic history to the great benefit of both subjects.

Now for Part I: Theory. Although it is hard to be sure, as I read it Pasinetti answers his titular question (J. M. Keynes's 'revolution' – the major event of twentieth-century economics?) in the affirmative. If so, I would agree. But this hardly matters, because what Pasinetti talks about is the content of the revolution and the extent of its success. He thinks of a revolution in Kuhnian terms, and considers the essence of the Keynesian revolution to be the attempt to create a monetary theory of production. Pasinetti contrasts the preanalytic vision of a pure exchange economy (Walras) with that of a pure production though nonmonetary economy (Smith, Ricardo) and argues that Keynes was trying to develop yet a third vision that assimilated money to a pure production economy. He considers the revolution to have had, at most, only a limited success and is unsure of whether it is possible to complete it successfully.

Pasinetti considers the reconciliation of GT with the Walrasian vision, even when that vision is expanded to permit unemployed resources and disequilibrium prices, to constitute an abandonment of the revolution. He feels that, however desirable, such an expansion of the Walrasian vision could have been accomplished without a revolution. On this I disagree, at least in part.

To begin with, Pasinetti underestimates Keynes's ambivalence toward the "revolution". No doubt, the idea of being in the company of Darwin, Einstein, Newton et al appealed to his vanity, and he appreciated the need for altering the mindset of contemporary economists and relished the role of prime mover. But what aspects of the mindset needed to be changed? Essentially, it was those aspects that bore upon the place of concepts of general levels of output and employment (unemployment) in the preanalytic vision of the exchange paradigm. Apparently it was not foreseen that the required changes could be accomplished within the framework of the exchange economy paradigm, but it turned out that it could. Once he perceived this, Keynes abandoned (what for him) was a secondary goal by (inter alia) endorsing Hicks's IS-LM apparatus as a translation of GT. This made it possible to achieve his primary goal, stated as follows (GT, vi): ". . . it is my fellow economists, not the general public, whom I must first convince".

On a strict interpretation of Kuhn's concept, such abandonment may have involved abandonment of the projected revolution. But so much the worse for the concept. In the half century subsequent to publication of GT, the Walrasian exchange paradigm has been drastically augmented – in good part to accommodate Keynesian ideas. This augmentation made it impossible, even after the neoclassical counterrevolution of the third quarter of the last century, for economists to exclude the question of effective demand from policy discussion. I consider that these two effects, combined, justify use of the term, "Keynesian Revolution".

To be sure these effects are not the revolution to which Pasinetti refers, and which Keynes and his quondam followers may have originally sought. That revolution was a revolution of thought which it was believed would be ancillary to a revolution of policy. When it turned out that such a revolution was unnecessary for its intended practical purpose – making the level of effective demand a key concept for discussion of economic policy – Keynes allowed it to lapse in desuetude, though (I suspect) traces of it remain in Chapter 17 of GT (in the discussion of own rates of interest) and in the later writings of Joan Robinson on capital theory and of Piero Sraffa on production of commodities. But given that in an augmented exchange (Walrasian) paradigm Say's Law will not be automatically satisfied, what is the practical justification for replacing it with a production paradigm as Pasinetti seems to desire, even if that could be done – a matter on which Pasinetti is unsure?

I can see some interesting theoretical puzzles arising from the replacement of one paradigm with the other, and each paradigm highlights a different aspect of a complex reality, but I do not see that such a paradigm shift is required to consider questions of inadequacy of effective demand or underemployment of resources. Such a shift might lead to a shift in views of the distribution of income and/or welfare, but that is another matter. But what made GT a revolutionary document, despite its frequent obscurity, is that it proposed a drastic change in the theoretical underpinning – paradigm shift or not – for a whole family of major policy decisions. This proposal was successful and, despite subsequent adjustments of doctrine, the change persists to the present.

Axel Leijonhufvud's essay (Mr. Keynes and the Moderns) contrasts (what he considers) two traditions in economic theorizing: the British Classical tradition in which he places Keynes and the "modern" tradition stemming from Walras which is expressed in contemporary general equilibrium theory. He presents the literature about GT as a series of attempts to reconcile the two traditions, paying due attention to his own contributions to this literature. As might be expected, the paper is well written and I found it stimulating. But while one can learn Leijonhufvud's opinion on a number of methodological issues from this paper, its brevity precludes serious argument on behalf of any of them.

Robert Skidelsky's paper (The conditions for the reinstatement of a Keynesian policy) is a beautiful counterweight to Pasinetti's piece. Pasinetti sees Keynes as a revolutionary gone astray; Skidelsky sees him as "cautious and hesitant" (p. 44) and reluctant to follow through on the implications of the revolution at which GT was seemingly aimed. (In my opinion, Keynes was a chameleon par excellence, both intellectually and personally; i.e. Pasinetti and Skidelsky are both right.)

In an initial section ("A little history") Skidelsky describes the undoing of the Keynesian revolution – both in theory and in policy – with evident relish. In the following section ("What would Keynes have said?") he presents the imagined responses of Keynes to "the main criticisms of the Keynesian system which had emerged by the early 1970s" (p. 40). The responses are those of a very moderate Keynes who would have amended the argument of GT to accommodate the criticisms of Friedman et al and proposed policies (for the 1960s) very different from those advocated by "hubristic" Keynesians such as Tobin or Galbraith.

In the final section (The case for, and conditions of, a Keynesian revival) Skidelsky presents his version of Keynesianism the flavor of which is conveyed by its concluding paragraph:

> the main condition of restoring even a modest version of Keynesianism – let us call it Eisenhower Keynesianism – is that budgets should be *balanceable* at low rates of inflation and unemployment. The main requirement for this is that developed-country governments should spend far less than they do – between 5 and 15 percentage points less, depending upon country and tradition. Keynesian stabilization policy is inconsistent with the continuous growth of public spending. You can have one or the other but not, beyond a certain point, both. The essence of effective Keynesian government is that it should be *limited* in scope and ambition. For all his backsliding Keynes recognized this. As Christopher Allsop has rightly written: 'The development of Welfare States, industrial intervention, and public expenditure programmes . . . has little to do with the economics of Keynes. It is necessary . . . not to lose sight of the fundamental point that the original message was minimalist in spirit' (p. 49).

Surely this view of Keynes deserves our respectful attention, as does the opposite view expressed by Pasinetti. However if viewed through the spectacles of either Skidelsky or Pasinetti, Keynes's impact on economics seems greatly diminished as compared with what appears through the bifocals of this reviewer.

Ruggles and Ruggles's MACRO- AND MICRODATA ANALYSES AND THEIR INTEGRATION

Kirsten K. Madden

By Nancy and Richard Ruggles, 1999, Edward Elgar Publishing Limited.

In 1952, with only thirteen years of publishing experience in quantitative economics, Richard Ruggles wrote:

> The plethora of data flooding the shelves and filing cabinets of economists is not necessarily a good omen of progress; it could mean that economics is heading for a statistical chaos wherein all data are internally inconsistent over time and incomparable with each other. Each economist, through proper selection, could then prove his own hypotheses and disprove those of his colleagues (p. 283).

Through the second half of the twentieth century, Richard and Nancy Ruggles worked extensively to "clean up" economic data files. Either Richard or Nancy Ruggles (or both) published work in economics in all but four years during the period 1939–1999. *Macro- and Microdata Analyses and Their Integration* (1999) reprints eighteen of their pieces from across the period, creating a sort of greatest hits anthology compiled from more than 130 works. The Ruggles' primary research interests reside in empirical matters: national income

Research in the History of Economic Thought and Methodology, Volume 19A, pages 219–228.
Copyright © 2001 by Elsevier Science B.V.
All rights of reproduction in any form reserved.
ISBN: 0-7623-0703-X

accounting and the creation and integration of datasets with micro- and macro-orientations. They also frequently used their empirical insights to create and critique theoretical constructs and economic methodology.

Although the Ruggles had noteworthy publishing careers in quantitative economics through six decades of the twentieth century, the decision to republish their work at this time raises questions. Who is the intended audience? Why would that audience be attracted to this particular collection? Why re-print a collection of the works of one (or two) contemporary academics? How does the work generally fit into economic thought, and to date, has their work made a substantial impact in the profession?

There are two potential audiences for this collection. Empirical economists actively engaged in the creation, integration or analysis of massive datasets may obtain substantial insights. Historians of economic thought interested in the twentieth century quantitative movement will also find this volume of interest. In particular, the Ruggles' work suggests a noteworthy gap in the current historical literature. Although the historical literature has recollected the history of econometric tools (e.g. for events in the first half of the twentieth century (see Morgan, 1990), concerning the history of structural estimation, see (Epstein, 1987), and for a history of time series analysis (see Klein, 1997)), historians of thought have not yet provided a general account of the creation, the use (or the mis-use) of the raw materials – the economic data – in empirical analysis. The importance of this topic and its relevance to the creation of economic knowledge stands out in a careful reading of the Ruggles' work.

Three primary reasons justify the re-print of a collection of the works of contemporary academics: (1) the collection is desirable because it has a known and substantial impact in economic thought; (2) the reprinted papers are difficult to obtain; or (3) the collection itself is a mechanism for introducing yet unrecognized contributions into collective disciplinary memory. While some empirical economists generally knowledgeable of the Ruggles' contributions may be interested for the first reason, the third is likely a primary rationale for this publication. The contributions of the Ruggles are not (yet) explicitly discussed in the historical literature about quantitative economic thought. The question is whether their contributions are substantial enough to warrant enshrinement by historians; three basic weaknesses of the current volume limit the ability to answer this question. First, the current volume provides scanty biographical content; we are not even told such basic information as where they attended school, for instance. Second, the economic and empirical content of this 514-page re-printed collection is remarkably repetitive. Although this kind of repetition is of interest to a historian tracing the transformation of a scholar's ideas over time, it can be a hindrance to the academician with a general interest in the topic but limited time. Third, the volume does not evaluate the actual impact of the Ruggles' research in economic thought and methodology. For example, it would be helpful to know whether changes the Ruggles recommended to national income accounting procedures were adopted, or whether their work building data systems at the National Bureau for Economic Research (NBER) made any substantial impact outside of that institution.

A brief biographical sketch (much of which is not available in the book) will serve to relate the course of the Ruggles' careers. Richard Ruggles was a Harvard graduate and worked for the Office of Strategic Services during World War II. Nancy Ruggles earned a Ph. D.from Radcliffe; the two married in 1946. From the mid–1940s on, Richard Ruggles' career centered around his professorship at Yale University, including a three year stint as department chair from 1959–1962. Nancy was hired by the Yale Economic Growth Center in the mid–1960s to design a data framework for country studies in economic development, but was asked to resign because of departmental concerns regarding employment of faculty wives. The Ruggles' other professional affiliations included the U.S. government (including the CIA), the United Nations, the Economic Cooperation Administration, the Ford Foundation, the Econometrics Society, the Agency for International Development, and the International Association for Research in Income and Wealth. Beginning in the late 1960s, the Ruggles' research agenda was put into action at the NBER. In 1978 Richard Ruggles moved his research back to Yale, at its Institution for Social and Policy Studies. At this time Nancy Ruggles was associated with the United Nations Statistical Office and followed her husband to the Yale Institution in 1980. In 1987, Nancy Ruggles was involved in a fatal accident. After 1990, Richard Ruggles resumed his writing and publishing activities, including a speaking engagement as a distinguished lecturer at the 1993 Allied Social Science Association meetings.

Thus, with little effort a biographical sketch can be created for the Ruggles suggesting that they had active and noteworthy careers in the profession. More troublesome in using the current volume to establish the Ruggles' intellectual contributions is its lengthy and repetitious nature. Although each of the reprinted chapters contain novel material and insights, the overall impact would be better served by a condensation of the material. The repetitions are exhibited in several ways. Tables are repeated across chapters. For example, in Chapter 18 tabulated information concerning household gross saving is the same as that provided in Chapter 4. Identically formatted tables can be found in Chapters 1 and 2; some of the numbers differ but there is no complete explanation for the change in values. Sections of papers are often quite similar in content. For instance, Section 1 of Chapter 4 is broadly similar to the section entitled "The

Theoretical and National Accounting Approaches" in Chapter 3. Some abstracts appear twice, once in the abstract summary in the beginning of the book and again at the beginning of some chapters (see Chapter 18 for an example.) Text is repeated word for word in Chapters 15 and 16 and in Chapters 8 and 14. Major and minor theoretical, empirical and methodological points are reiterated throughout the papers, but at the same time the reader cannot be encouraged to skip many chapters because new and relevant insights would be missed. Fear of missing the new coupled with repetition-induced boredom makes for a particularly frustrating read. Likely only a historian of the Ruggles' economic thought and methodology will truly appreciate the collection as a vehicle to analyze their economic thought.

The disciplinary impact of the Ruggles is a trickier issue to analyze. In what follows, I outline some of the Ruggles' major contributions. Assuming as the publisher does, that their research agenda warrants attention, I make suggestions for filling a gap in the history and methodology of quantitative economic thought.

The Ruggles' macroeconomic writing falls squarely into the Keynesian school with their conceptual focus on aggregate expenditure, consumption and investment. They encourage government intervention to avert economic calamity. For instance they recommend the government as employer of last resort in economic downturns, with the caveat that employment must "contribute in an important way to the output and productivity of the economy" (p. 54). They also reflect institutionalist thinking. A fundamental criticism of national income accounting procedures is the lack of attribution of economic transactions to the appropriate groups of decision makers. They argue that consumption and investment decisions are made both by enterprises and by households and that the national income accounts should reflect these institutional realities (e.g. see Chapters 2, 3, 18). In the articles reprinted from the mid–1970s, the Ruggles also exhibit relatively progressive thinking for the period in their desire to nationally account for the environmental impact of economic activity and for the non-traded production of housewives (see Chapters 15, 16).

The non-quantitatively inclined researcher will find this volume accessible. Given their substantial expertise in empirical matters, it is interesting to note that the Ruggles do not present their results in mathematical form; they state their equations verbally and present their results in simple tables of statistics. Explicit mathematical expressions are only presented in a paper analyzing fertility. In "Methodological Developments" we find noteworthy criticism of the mathematical approach in economics, as well as justification for their own research and presentation style: "Much of the most effective statistical analysis is of this extremely unelaborate type, bringing pertinent information to bear on specific problems by presenting the statistical data which are available" (pp. 282–283).

Three general topics are highlighted in this collection: macroeconomic analysis, microeconomic analysis, and macro- and microdata integration. Five articles spanning 1975–1993 summarize their contributions to macroanalysis. This section makes the clearest links between data and theory. With a Keynesian theoretical framework in mind, a general theme of this section is to "inject some empirical reality into saving and investment theory to help conserve the vitality of the subject" (p. 57).

The Ruggles' novel macroeconomic contribution is their critique of national income accounts classifications of economic data. For example, the consumption accounts are criticized for including non-household directed elements such as nonprofit institutions and pension contributions by firms. The investment accounts are criticized for including owner-occupied housing purchases. The Ruggles argue that theoretical models inappropriately link consumers with consumption and saving, investors with production and investing; in reality, all sectors are involved in these economic activities. Since investment and saving activity are found in all sectors of the economy, the Ruggles argue that each of the components of aggregate expenditure should include savings and investment accounts. Accordingly, the Ruggles adjust national income accounts figures for the period 1949–1973. Their revised figures suggest that in aggregate, households generally borrow when purchasing homes and lend otherwise. Since the majority of household saving is institutionalized (e.g. pension contributions, mortgage contracts, etc.), they hypothesize that consumption is the residual element fluctuating with output rather than saving. In prosperous times, the household sector borrows because it invests more than it saves when purchasing homes and other large consumer durables; in recessions, household saving tends to exceed household investment and so this sector becomes a net lender. In aggregate, the household sector is not a major supplier of savings for other economic sectors because in "effect, all the household sector is managing to do is to provide almost enough saving to cover its own purchases of durables" (p. 34). Insurance and pension reserves are the main source of intersectoral lending.

Enterprise data is disaggregated into industry groups, and enterprise gross saving and investment are compared to determine net lending and borrowing sectors. Generally within the enterprise sector, self-financing of investment activity is the norm. In the Ruggles' view, economists lamenting low levels of consumer saving are misguided because (with exceptions) the primary source of investment financing comes from retained earnings of the firm, not from the consumer: "the general rule is that each tub tends to stand on its own bottom" (p. 117). In the final paper in this section, this conclusion is slightly altered in noting that although gross saving approximates gross capital formation in each sector, this "does not mean that there is no intersectoral financing" (p. 132).

Lacking capital accounts for the government in the earlier macroeconomics papers, the Ruggles review budget trend data. In considering the total impact of all the components of aggregate expenditure on output, the Ruggles argue that output does not typically reach full employment levels because aggregate savings tends to be greater than investment in recoveries. An important theme falling out of their macroanalysis is that

> the simplistic and abstract theories of saving and investment are not an adequate basis for empirical measurement and . . . empirical measurement without theoretical content is not meaningful. Theoretical structures need to be able to take into account the institutional characteristics of economic systems rather than to assume they do not exist. Empirical measurement, in turn, must be based on operationally meaningful economic concepts (p. 72).

Macroeconomic data should be reformulated to obtain full "sets of income accounts, capital accounts and balance sheets for different sectors and subsectors of the economy" (p. 136).

Five microeconomic articles spanning the period 1947–1984 highlight the Ruggles' contributions to microanalysis. The first two articles in this section (1947, 1960) fit the collection only because of the extremely broad nature of the title. (How many economics papers couldn't arguably fit into a volume entitled *Macro- and Microdata Analyses and Their Integration*?) The papers on armament factory markings and fertility analysis have no clear connection to the other articles in the volume, and the reader short on time may skip these 50 pages. The two papers contributing directly to the general theme of the book concern strategies for merging microdata sets. Writing in the 1970s, the Ruggles witnessed a substantial growth in microeconomic databases. One of the problems they observed is that researchers limited to a single database can not necessarily access all the variables required for a complete analysis. Researchers devised methods to address this dataset limitation. The Ruggles review existing techniques for merging microdata sets using exact matching and statistical matching procedures and describe the limitations of each. They present their own statistical merging method and provide an empirical illustration of its application, including a test of the validity of the merging method.

Interestingly, in "The Role of Microdata in the National Economic Accounts," the Ruggles briefly describe an idea to create all-purpose microdata sets with "no real microunits" (p. 349). They hypothesize mapping known information concerning population distributions to an artificially constructed data set. For example, knowing the age distribution, sex distribution, and income distribution in a given population, it is possible to create an artificial household dataset by mapping these characteristics to each observation in the appropriate population distributions; similarly population characteristics could be distributed to a synthetic dataset for firms and the government. It is unclear how seriously they toyed with this idea as they state that in "a large country, the number of local government units is very substantial, and postulating their existence and the general nature of their activities even on the basis of the most casual empiricism *is not as ridiculous as it would appear*" (pp. 348–349). As new information is obtained from real data the synthetic data set would be adjusted accordingly and it will "thus correspond more and more closely, statistically, to the information content of real microdata sets, even though at the microunit level they still contain no real microunits" (p. 349).

The final paper in the microanalysis section summarizes the analytical uses obtained from a newly created longitudinal establishment data set (LED). In particular, the Ruggles find that, given the variation in firm behavior, the theoretically specified "representative firm" in microeconomics really only represents a small proportion of firms. A major strength of the Ruggles' work concerns their knowledge of the weaknesses of datasets and their ability to draw appropriate conclusions regarding the reliability of results generated from the data. For example, their intimate knowledge of data imputation procedures in the LED file allows the Ruggles to couch the reliability of their conclusions in the context of firm size: the "LED file is not as well suited for analyzing small establishments as large ... the conclusions developed for small establishments, and the comparisons between small and large establishments, should be viewed with caution" (p. 241).

Another noteworthy feature of the Ruggles' work is their ability to iterate across the micro- and macro-analytical frameworks. The iteration between the micro- and macro-contexts allows them to develop theoretical explanations of data results that make more sense. In their microeconomic analyses, they typically have macroeconomic issues looming on their research horizon. For example, firm decision making may be couched in terms of the general state of the economy in which the decision takes place. In the macroeconomic analyses, their theoretical vision of decision making by the microunits also shape their results.

Eight articles spanning the period 1952–1987 introduce their contributions integrating macro- and microeconomic datasets. The link between technology, methodology and theory hypothesized in "The Relation of Methodology to the Technology of Economic Research," provides a general theme for a new treatise on the quantitative movement in economic thought. The Ruggles argue that technological innovations in data accumulation, storage and processing have major implications for empirical methodology, and through methodology spill over into economic theory. To make their point, they provide a brief history of technology along with the impact in economic methodology. For instance, before substantial technological developments, raw observations of Census data could not be shared conveniently. With the application of adding machines at the end of the nineteenth century, data summaries became more feasible and data was reported in aggregated form. With the innovation of punchcard technology, data cross-tabulations became the norm. The computer revolution of the late twentieth century had major implications for data storage and access. With the development of magnetic tapes for data storage, it became generally feasible for researchers to access raw data instead of data cross-tabulation summaries. Computing changes also promoted the general use of advanced statistical manipulation of data. The revolution in computing technology is recognized by the Ruggles as necessitating a revolution in scientific practice:

> But it does require a change in attitude and emphasis. The new technology . . . has made possible a wholly new conception of the statistical function and, to take maximum advantage of the potentialities, a rather radical change in modes of thinking is required. Where statisticians were . . . accustomed to planning their work in terms of the production of a set of prespecified tabulations, it is now necessary for them to reorient their thinking to deal with a new set of problems and a new set of goals. . . . The new tools have enormous potential for application to increasingly pressing social and economic problems (p. 448).

To do empirical analysis well the Ruggles argue that techniques are needed to merge various microdata sets and to provide better integration of micromodels into the macrodata accounting system.

The Ruggles' major goal in micro- and macrodata integration is to devise a general data system which is fully integrated across economic sectors. A primary objective is to encourage consistency such that microdata are linked to macroeconomic accounts and the microdata totals sum up to macroeconomic aggregate quantities. This general recommendation reflects years of frustration with the lack of consistency across data collecting institutions.

To reach their goal, the Ruggles originally argued for the use of the national income accounting framework with four basic sectors: enterprises, households, the government and an external account. Data for each sector is summarized in income and outlay accounts, savings and investment accounts and balance sheets. The sectors themselves are capable of sub-classification (e.g. enterprises might be broken down into corporate, nonprofit, agriculture, etc.) Demographic and social data can also be integrated into these accounts at the micro-level. For instance, in an extensive team project at the NBER, researchers working on a large data integration project with the Ruggles were responsible for introducing environmental data and undertaking time allocation studies to begin estimation of non-market household production. The general idea is to adjust both micro- and macrodata accounting to facilitate fully consistent integration across datasets. Microdata is to be integrated with the national accounts "so that the microdata sets which are developed become identifiable parts of a compre- hensive economic and social information system" and national accounts are revised so sectors directly refer to reporting units in the economy, the reporting units being consistent across the datasets (p. 335). The Ruggles do not advocate throwing out the existing national accounts system, but rather "a modest realignment of the sectoring of existing national accounts to distinguish different types of reporting units, so that the accounts can provide the control totals for specific microdata sets" (p. 336).

In some of their writing, the Ruggles carry their conceptualization of an integrated data system to an extreme by suggesting a data clearing house in the form of a Central Statistical Office. Obvious concerns regarding confidentiality and personal privacy are only noted briefly in the volume. For example, in "The Development of Integrated Data Bases for Social, Economic and Demographic Statistics" Nancy Ruggles devotes two pages to confidentiality and privacy issues, including a brief discussion of methods to facilitate confidentiality. It appears from the book's bibliography that the Ruggles did address confidential- ity issues in numerous research papers, but space constraints limited their reprint in this volume.

The researcher that takes the time to carefully read the 514 pages in this volume is likely to conclude that the Ruggles' intellectual contributions are substantial. The question is whether economists who have limited (or no) familiarity with the Ruggles' work will see the payoff in weeding through so much repetition to independently summarize the Ruggles' main contributions. As Richard Ruggles points out in his methodological article, the discipline highlights "theoretical development of hypotheses" over other stages of research. Since the Ruggles' primary contributions do not lie in this stage, the mode of delivery chosen to introduce economists to their work should have been considered more carefully. Standing on its own, this anthology is not likely to work as a mechanism to solidify the Ruggles' contributions into disciplinary memory.

In 1990, Morgan published the clearest account of the early history of the econometrics movement, making it apparent that the discipline would benefit by having convenient access to a collection of some of the more influential contributions to econometric history. Morgan and Hendry (1995) provides this collection, including an 80-page introduction to the material. This is an effective approach to introducing new material into the history of thought. Likewise, the Ruggles' contributions would be more completely digested and integrated into the history of economic thought if a condensed summary of their contributions were provided in 100–200 pages, including an extensive biographical overview and a clear account of the impact their work had in the discipline. A more ambitious historical research project would be to trace the events associated with the collection, organization and dissemination of the raw materials – the data – used in the quantitative analysis of economic phenomena and to analyze the influence on methodology and the creation of economic knowledge. The Ruggles' work could justifiably be highlighted as a substantial contribution in such a treatise. The central role they attribute to technological change in the collection, storage, and dissemination of economic data would merit attention. Work in the philosophy of science by Pickering (1997) could provide a theoretical underpinning for such an account.

REFERENCES

Epstein, R. J. (1987). *A History of Econometrics*. Amsterdam: North-Holland.

Hendry, D., & Morgan, M. (1995). *The Foundations of Econometric Analysis*. Great Britain: Cambridge University Press.

Klein, J. (1997). *Statistical Visions in Time: A History of Time Series Analysis, 1662–1938*. Great Britain: Cambridge University Press.

Morgan, M. S. (1990). *The History of Econometric Ideas*. Great Britain: Cambridge University Press.

Pickering, A. (1995). *The Mangle of Practice: Time, Agency, and Science*. Chicago: University of Chicago Press.

Ruggles, R. (2000). The Yale Economics Department: Memories and Musings of Past Leaders. http://www.econ.yale.edu/depthistory.html. Accessed May.

Perlman and McCann's THE PILLARS OF ECONOMIC UNDERSTANDING: IDEAS AND TRADITIONS

LEGACIES AND TRADITIONS IN ECONOMICS: A REVIEW ESSAY

Anthony Brewer

University of Michigan Press, 1998, xx + 639 pp.

At first sight, one might easily take this for another retelling of the old story of the history of economics from the mercantilists and Petty to the twentieth century. It is certainly not that, though it does cover most of the usual names (plus a few not usually considered important). It is harder to say exactly what it is. The authors promise to pick out 'grand ideas and central themes or paradigms of economics' which have been widely ignored, and to examine basic questions which 'concern such extraeconomic issues as Natural Law and natural rights, property rights, the design of social structures, and even philosophical questions regarding the basic elements of human nature'. There is to be a companion volume, entitled *Factors and Markets*, but there is relatively little in this volume to explain the division of labour between the two.

I can perhaps best convey what is *not* dealt with in this volume by giving a couple of examples. Ricardo gets only a short section, mainly focused on his

Research in the History of Economic Thought and Methodology, Volume 19A, pages 229–239.
ISBN: 0-7623-0703-X

pamphlet on representative government, with nothing of substance about the labour theory of value, rent or profit, while the rather longer section on Böhm Bawerk stops short of discussing his capital theory, reserved for the companion volume. Clearly, much of what one would expect to find in a fat book on the history of economics is not included in this one.

What, then, is the book about? The clearest short statement I managed to find appears almost in passing in the section on von Mises. 'In line with the theme of the present work, we wish to review Mises's contribution to the understanding of the place of the individual versus the community, and the role of economics as a discipline' (p. 443). Perlman and McCann relate their treatment of these issues to what they call 'patristic legacies', which I shall discuss at the end of this essay.

THE INDIVIDUAL AND THE COMMUNITY

The most prominent theme of the book is the history of 'competing views of human nature and of the institutional structures for the promotion of social welfare' (p. 233). To show what is involved, let me take utilitarianism as an example. (The example is mine, and is simplified to the point of caricature, though much of it is distilled from the book.) Think first of the treatment of individuals. The simplest form of utilitarianism treats the individual as a simple maximizer of utility, defined as pleasure minus pain. Modern economists typically empty out the content by defining utility as whatever the individual maximizes, but many nineteenth century economists wanted to put more content in, or vary the content. Are pleasure and pain to be considered in physical terms? Are some pleasures better ('higher') than others, and is there a place for education in teaching people what they ought to enjoy? How can morality be incorporated into the story? How could utility be measured or observed?

Now consider utilitarianism as a theory of social welfare. In its crudest form, it would tell us to maximize the sum of utilities. Taken literally, that implies that the welfare of any given individual has no importance except as a component of the total. A consistent utilitarian would be willing to inflict the most horrific suffering on an individual if the gain to others outweighed it. There is, on the face of it, no space for moral concerns or for any concept of the rights of individuals in the story. Can they be incorporated in some way? Should we think of the utilitarian principle as a guide to specific choices, or as a higher order principle used to construct rules which have to be followed even if they lead to a loss of utility in particular cases? If the latter, can utilitarianism provide a justification for a concept of individual rights, perhaps because a system which assigns certain rights to individuals will have a higher sum of utilities than one which does not, or do concepts of individual rights and other moral concerns have a justification independent of, and perhaps conflicting with, utilitarianism?

Utilitarianism is only an example, though it does bulk fairly large in Perlman and McCann's discussion. There are plenty of non-utilitarian answers to similar questions. It is argued here, for example, that mercantilism was characterized by an overriding concern with the power of the nation state, an aim defined at the social (national) level independent of the utilities or happiness of the individual citizens. Again, private property rights are an important aspect of 'the institutional structures for the promotion of social welfare'. Locke and others thought that property rights were a matter of natural law, independent of any actual society or legal framework. From this point of view, legal concepts of property must be founded on natural law and should be criticized if they deviate from it. Others have argued that property is a wholly social concept, to be justified and, if necessary, adjusted on utilitarian or other consequentialist grounds. Different economists have offered different answers (explicit or implicit) to questions of this sort. Tracing the twists and turns of the story is a fascinating theme.

RATIONALISM VERSUS EMPIRICISM

A second theme in the book is the methodology of economics, seen mainly in terms of a recurring contrast between Cartesian rationalism and Baconian induction. The authors trace a Baconian strand in the history of economics, running in its purest form from Petty and his successors in political arithmetic to the founding of the Econometric Society, with an opposing Cartesian tradition which emphasises pure theory and logical coherence over empirical observation. Von Mises' description of economics as an a priori science based on intuitively certain and unquestionable premises is a notable exhibit.

This methodological polarity between empiricism and rationalism raises problems which Perlman and McCann's other themes do not. In the case, say, of the relation between the individual and society, it is clear that what is under discussion is simply what different authors have said about the subject. Methodology is not so simple. One of the lessons of modern discussions of the philosophy and sociology of science is that publicly announced methodological principles may not be a good description of what is really done. As McCloskey puts it, 'economists do not follow the laws of enquiry their methodologies lay down' (1983, p. 482). In any case, neither Baconian inductivism nor Cartesian rationalism really stands up on its own. Baconian data collection can only proceed in a framework that defines what is to be measured and how. Take Petty as an example. He did not collect much data himself, nor was he able to draw any very surprising conclusions from the sparse data at his disposal. His real achievement was to construct a framework to fit the data into. Cartesian rationalism is equally unable to stand on its own. Logic alone gets nowhere without premises to start from, and 'cogito ergo sum', the only thing Descartes thought he could say for sure, did not get him far. Descartes could no more exclude empirical evidence than Bacon could rely on unadorned facts.

Perlman and McCann do not explain the criteria they use to assign different authors to one camp or the other explicitly (they rarely set out definitions or concepts explicitly), but the focus is clearly on what different authors said they were doing, or on visible characteristics of their style. Thus, the political arithmeticians are assigned to the Baconian camp because Petty clearly thought he was following Bacon, and the others followed Petty, but overall the treatment of methodology is disappointingly thin. John Stuart Mill on method, for example, is surely worth at least a little discussion (see Hausman, 1992, pp. 123–133) but is passed over in silence. In defence of the book, one might say that serious study of the history of methodology is in its infancy and that to disentangle the real methodological issues over a range as wide as that covered here is probably impossible.

SCHOOLS OF THOUGHT AND NATIONAL TRADITIONS

This is a long book, clearly the product of an enormous amount of reading and thought. The bulk of it is devoted to a survey of the history of economics from the beginnings to the twentieth century, with the emphasis on the range of issues described above. After an introduction and a chapter on the British patristic legacy (see below), come three chapters on schools of thought which lie somewhat outside the main line of development: mercantilism, political arithmetic and econometrics. The chapter on mercantilism emphasizes the rise of the nation state and state power as an aim of policy. The chapter on 'the measurement of economic magnitudes', starts with Petty and political arithmetic and includes (rather oddly) Cantillon and the Physiocrats, while the following chapter jumps to the twentieth century to discuss the origins of econometrics. The remainder of the book (pp. 223–568) is divided by national schools. There are three chapters on Britain – on the classics, the neoclassicals, and a chapter divided between the British historical school and the 'post-Marshallian Cambridge tradition' (Pigou, Keynes and Joan Robinson). The German and Austrian tradition gets a chapter, as do the French (allegedly Cartesian) tradition and American institutionalism.

The British tradition plays a central role, but the way it is presented may come as a surprise. The chapter on the classics is subtitled 'individualism, utilitarianism, and property rights', while the chapter on the neoclassicals is headed 'utilitarianism as the basis for "scientific" economics', and the chapter on post-Marshallian British economics is subtitled 'a reinterpretation of utilitarianism'. The classical tradition is seen as primarily, if not quite consistently, utilitarian, and as leading directly to the British neoclassicals. Smith appears alongside Hume and Hutcheson as a rather distant precursor, with Bentham as the central founding figure and James Mill as his main follower. Ricardo was 'steeped in the Utilitarianism of Bentham and James Mill' (p. 264). John Stuart Mill's *On Liberty* and *Utilitarianism* are the culmination of the classical line and remained influential until the end of the nineteenth century. Pigou's welfare economics continued the tradition but Keynes stood somewhat outside the utilitarian tradition because of his doubts about the possibility of rational calculation in an uncertain world. This is very different from the way the story is usually told, but seems reasonable enough given the focus of the book. The discussion of different varieties of utilitarianism and their development is one of the strongest features of the book.

PATRISTIC LEGACIES

Perlman and McCann draw the different threads of their history together by identifying a number of 'patristic legacies'. My dictionary defines patristic as 'of (the study or writings of) the Fathers of the Church', but Perlman and McCann have coined a secular adaptation of the term. 'Patristic approaches . . . are basically the approaches . . . recognizing an appeal to authority, to those figures accepted as fulfilling a paternalistic role' (p. 3). They recognize that few modern scholars will admit to any reverence for the wisdom of the past, but claim that 'we do have patristic elements almost indelibly etched on our minds' (p. 3), albeit a hotchpotch of different elements derived from different sources. There seem to be several elements to this claim, which is never set out very clearly. First, it is claimed that there are perennial themes in the history of economics, which recur again and again, such as the relation between individual and society. Second, it is claimed that we come to these issues with a mind set formed, consciously or unconsciously, by our predecessors. Third, major traditions can be identified with specific 'patristic' figures, founding fathers whose ideas live on.

The introductory chapter deals with some ancient and medieval figures (Plato, Aristotle, Aquinas), but they play a relatively little role in what follows.

The patristic traditions which are the main subject of the book emerge as the argument proceeds but they are not always clearly defined or signalled. It is best at this stage to turn to the concluding chapter, where the main themes are set out explicitly and traced back to their 'patristic' origins (pp. 571–578). The relation between the individual and the community is seen as an opposition between 'communitarianism' and individualism. The key figures here are Hobbes, Locke, and Rousseau. Hobbes is identified with a statist view in which men are 'too selfish or brutish to live communally' (p. 571) without a central authority imposed by force, Locke is seen as the defender of the rights of individuals, and Rousseau is identified with a utopian communitarianism based on man's natural goodness. A linked theme is the right to private property, asserted by Locke but less important in 'the Rousseau-influenced sphere' (p. 575). The methodological opposition between empiricism and rationalism is a second theme, presented as a conflict between the patristic figures of Bacon and Descartes. Further themes include differing views of the role of work, which derive from biblical sources and surface in Locke's justification of property and in the Ricardo/Marx labor theory of value, and of the 'preferred way of economic life'. where Perlman and McCann describe a shift of focus from agriculture (in physiocratic theory) to manufacturing (in Marx, for example) as the industrial revolution proceeded.

Of the key patristic authority-figures listed here, four (Hobbes, Locke, Bacon, and Descartes) were active in the seventeenth century, one (Rousseau) in the eighteenth. All would figure prominently in any history of Western philosophy or in any discussion of the philosophical background to the enlightenment. The central message of the book under review is that economics has developed in the shadow of seventeenth-century debates on the relation between individuals and the state and on scientific method. As Perlman and McCann say of the Hobbes/Locke/Rousseau trilogy: 'later writers simply built on these designs' (p. 572).

There is a sharp contrast here with most accounts of the history of economics, but that is because Perlman and McCann focus on a different set of issues. The history of what Schumpeter called economic analysis is cumulative, in the sense that later writers build on the achievement of those who went before. By most reasonable standards it is a history of progress. The history of views of human nature and of social welfare, on the other hand, does not show the same sort of progress. One could reasonably argue that the same ideas, combined in different ways and adapted to changing circumstances, come back again and again. That is what Perlman and McCann imply when they emphasize the lasting importance of cultural and intellectual legacies going back to the seventeenth century. I will argue that there are problems with the way that Perlman and McCann state their case, but it is a case that is well worth thinking about.

CONTEXT AND CHRONOLOGY IN THE HISTORY OF IDEAS

There seems to me to be a contrast between Perlman and McCann's careful and detailed treatment of economics and their relatively casual approach to the history of philosophical and social thought. Economists are set in the context of their time, but philosophical ideas are not. If (say) a nineteenth-century writer puts the interests of the state above the rights of the individual, Perlman and McCann tend to apply the label 'Hobbesian' without further discussion, as though Hobbes were a disembodied spirit standing outside time. They rarely ask what it would mean to be 'Hobbesian' two centuries after Hobbes. This is a rather old-fashioned way of treating the history of ideas, reminiscent of Quentin Skinner's account of the old orthodoxy in the history of political thought. 'A canon of leading texts was widely regarded . . . as the only proper object of research. . . . The reason, it was urged, is that such texts can by definition be expected to address a set of perennial questions' (1998, pp. 101–102). Skinner has argued cogently that seventeenth-century writers were dealing with issues specific to their own time, with a conceptual framework and a set of presuppositions quite different from those of today. If that point is accepted, the idea of an unchanging set of 'patristic' legacies becomes harder to sustain.

One symptom of this problem in the text under review is a casual way of treating chronology when dealing with the philosophical background. I will give two examples. First, we are told that 'ecclesiastical responses . . . initiated the seventeenth century debate' about Hobbes (p. 49). Locke is treated as following on *after* these ecclesiastical responses, but in the order of presentation adopted in the book the first 'ecclesiastical' response to be discussed, placed ahead of the discussion of Locke, is by Joseph Bishop Butler, who had not even been born when Locke wrote his *Two Treatises on Government*. Also placed before Locke, as one of the responses which 'initiated' the debate, is Anthony Ashley Cooper, who was a child, and actually being tutored by Locke himself, when Locke was writing the *Two Treatises*. The text does not quite say that Butler and Cooper came before Locke, but the unwary reader might well come away with that impression.

Again, Perlman and McCann claim that 'to comprehend Physiocracy, one needs first to understand the philosophy of Jean-Jacques Rousseau' (p. 174), and they add, in their conclusion, that the physiocrats were 'in the Rouseauian mold (understandable given the French reverence for Rousseau)' (p. 572). They give not a hint that Rousseau and Quesnay were contemporaries (and that Quesnay was the older). They rightly identify Rousseau's *Du Contrat Social* of 1762 as a key work, but fail to note that Quesnay's framework was well established by about 1760 and could not have been influenced by it. The 'French reverence for Rousseau' could hardly have been at work at a time when Rousseau was in exile from France, fearing for his own safety. Influence there may have been, but it must be argued in terms of specific contacts between Quesnay and Rousseau and linked to the precise chronology of their work. What is happening here, I think, is that Perlman and McCann see philosophical debates in an abstract way, outside the mundane chronology which rules their discussion of the history of economics.

There is a related problem. Throughout the book, patristic legacies are identified with the names of specific authors (Hobbes, Locke, and so on). The very name 'patristic' ('of the fathers of the church') suggests an identification with a particular individual and so does the way the notion is introduced at the start of the book. At the same time, Perlman and McCann recognize that patristic legacies must often work unconsciously and indirectly. They become part of the furniture of our minds. This is clearly true, but it makes the relation between the named individual and the patristic tradition named after him somewhat unclear. I will take Hobbes as an example, because he plays a central role in Perlman and McCann's story and because there are particular difficulties in the way they deal with him. If some economic writer is said to be 'Hobbesian', does this mean that his or her work was (directly or indirectly) influenced by Hobbes or simply that it exemplifies views similar to those of Hobbes? Which of Hobbes's opinions are to be counted as part of the tradition labelled with his name? What is the relationship between a specific author, who wrote on a variety of subjects in a specific period and context, and the tradition ascribed to him? Perlman and McCann never confront these questions.

They rate Hobbes's influence very highly.

> The Hobbes Patristic Legacy, seen both in the negative and the positive senses, is almost without parallel as an influence in the development of economic thought in the English-speaking world. On the negative side it stimulated a variety of antitheses, from which modern constitutional government as well as modern economics have derived their justifications; and on the positive side it incarnated the foundations of scientific empiricism (p. 133).

This claim can only make sense if the 'Hobbes Patristic Legacy' refers to the intellectual legacy of a particular individual, Thomas Hobbes, since a unified state (which is what the negative legacy is about) and 'scientific empiricism' (the positive side) have no logical connection but are lumped together because, and only because, Hobbes advocated both. On the very same page, however, it is claimed that 'the mercantilists very clearly held to a Hobbesian view of the duties of the individual in relation to the good of the state'. This cannot refer to any direct or indirect influence of Hobbes as an individual, because most of the mercantilists under discussion wrote before Hobbes. Here the label 'Hobbesian' must refer to a view in which the interests of the state take priority over those of individuals, regardless of its source.

I would not dispute Hobbes's importance in the history of political thought. It is true that his support for an absolutist state was by no means unusual at the time, and that there are precursors for elements in his argument – for example, Calvinist writers had argued that a strong state was needed to restrain the appetites of fallen men – but he produced a strictly secular and (as he saw it) scientific argument at a level of philosophical seriousness that surpassed anything that had gone before. That said, Perlman and McCann's reasons for giving him patristic status need, at the least, to be stated with considerable caution. They seem to treat support for a strong state as, in itself, 'Hobbesian', but support for absolutism and other forms of statism comes in many varieties at different dates, before and after Hobbes, and was not necessarily, or even usually, based on his arguments.

Hobbes's negative influence is harder to show than Perlman and McCann admit. Locke's *Two Treatises of Government*, which they treat as a response to Hobbes, were ostensibly aimed at Filmer, not at Hobbes. Filmer was far more influential than Hobbes at the time Locke was writing, and may indeed have been the real target (Laslett, 1988, pp. 67–92). After all, *Leviathan* was very much a product of the civil war and Locke was writing in a quite different context some thirty years on. In addition, Hobbes's relation to 'scientific empiricism' is less straightforward than Perlman and McCann admit. He regarded geometry as the ideal science precisely because it was logically certain (Jesseph, 1996; Sorell, 1988, p. 71) and mistrusted experimental methods (Tuck, 1989, p. 49), so to put him in the Baconian camp against Cartesian rationalism is to oversimplify. I find the claim (p. 153) that the 'Hobbesian influence' led Petty to emphasize data over theory incomprehensible. If *Leviathan* is not a work of theory, what is? Finally, in treating Hobbes as a founder of what they call 'communitarianism' in opposition to Locke's 'individualism', Perlman and McCann should at least take account of claims that Hobbes was himself an individualist (see, for example, Ryan, 1988 and references given there).

My criticisms do not, I think, nullify all of Perlman and McCann's insights. Hobbes really was an important influence in the long run, though his influence was not as direct or as simple as they suggest. If Locke did not write as a direct response to Hobbes, Hobbes was still an important part of the background against which he wrote. Hobbes was an important contributor to debates on scientific method, even though he cannot be seen in any simple way as siding with Bacon against Descartes.

The way forward might be to replace the idea of an unchanging set of patristic legacies, each tied to an individual patristic founder, with some more flexible notion of a range of evolving traditions, allowing one to recognize that there are lasting contrasts, say between individualist and statist traditions, but that those traditions develop over time. Writers like Hobbes and Locke do have a continuing influence, but different generations draw different lessons from them and their individual contributions become submerged in a wider stream of thought. A subtle and contextual history of economics needs a subtle and contextual treatment of the intellectual background.

CONCLUSION

I found this a fascinating book, if also at times an infuriating one. It is not for the fainthearted or for beginners, but those who have the patience to work through it will find a genuinely original view of the history of economics. If I have criticized some of Perlman and McCann's arguments, it is because they pose important issues which deserve further discussion.

REFERENCES

Hausman, D. (1992). *The Inexact and Separate Science of Economics*. Cambridge University Press.
Jesseph, D. (1996). Hobbes and the method of natural science. In: T. Sorrell (Ed.), *The Cambridge Companion to Hobbes*. Cambridge University Press.
Laslett, P. (1988). Introduction. In: J. Locke (Ed.), *Two Treatises on Government* (student edition). Cambridge University Press.
McCloskey, D. (1983). The rhetoric of economics. *Journal of Economic Literature, 21*, 481–517.
Ryan, A. (1988). Hobbes and individualism. In: G. Rodgers & A. Ryan, *Perspectives on Thomas Hobbes* (pp. 81–105). Oxford: Clarendon Press.
Skinner, Q. (1988). *Liberty before Liberalism*. Cambridge University Press.

Sorrell, T. (1988). The science in Hobbes's politics. In: G. Rodgers & A. Ryan, *Perspectives on Thomas Hobbes* (pp. 67–80). Oxford: Clarendon Press.
Tuck, R. (1989). *Hobbes*. Oxford University Press.

Martin's VERSTEHEN: THE USES OF UNDERSTANDING IN SOCIAL SCIENCE

Tyler Cowen

New Brunswick, New Jersey, Transaction Publishers, 2000, pp. 264.

This readable and reasonable book is the definitive study of its topic. Martin examines the Verstehen tradition in the social sciences, ranging from Weber to Collingwood to Winch to Schutz to Geertz to Charles Taylor, among others. The survey is clear and focused.

Verstehen is a German word meaning "understanding", but has since come to acquire a more specific connotation of interpretative social science with a subjectivist slant. The Verstehen tradition is out of favor in neoclassical economics these days, but it has a long and distinguished lineage. One cited author, W. H. Dray, describes his conception of Verstehen as follows: "[the historian] must *penetrate* behind appearances, achieve *insight* into the situation, *identify* himself sympathetically with the protagonists, *project* himself imaginatively into the situation. He must *revive, re-enact, re-think, re-experience*, the hopes, plans, desires, views, intentions, etc., of those he seeks to understand". (p. 106). The social scientist seeks to comprehend the subjective mental states of the actors involved in a particular historical or

Research in the History of Economic Thought and Methodology, Volume 19A,
pages 241–244.
Copyright © 2001 by Elsevier Science B.V.
All rights of reproduction in any form reserved.
ISBN: 0-7623-0703-X

economic event. Clifford Geertz's account of a Balinese cockfight, published in his *Interpretation of Cultures*, is perhaps the best known example of the Verstehen method. Max Weber is the most prominent defender of the Verstehen method in the history of ideas.

The Verstehen tradition traditionally has been opposed to positivistic social science, with its evidence on quantitative laws and testable predictions. The Verstehen tradition perhaps is closest to the Austrian school of economics, and to some of the institutionalists, and has been cited as such by some of the Austrians, most prominently Donald C. Lavoie. As Lavoie points out, the writings of Mises, Hayek, and Kirzner can all be understood as trying to interpret economic phenomena in terms of the subjective mental states of individuals.

Martin surveys the different understandings of Verstehen offered in the literature and considers the strengths and weaknesses of each. His treatment is fair, easy to follow, and philosophically sophisticated. At the same time, he surveys the leading critics of Verstehen, including Hempel, Nagel, and Abel. These critics charge that the Verstehen method does not serve as a means of verifying theories and does not help us come up with useful statements of law-like behavior.

A typical chapter of the book takes one strand of the Verstehen tradition, outlines it, and examines its central claims. We have chapters such as "Verstehen and Ordinary Language Philosophy", "Verstehen and Situational Logic", "Verstehen and Phenomenology", "Verstehen and the Sciences of Man", "Verstehen and Anthropology", and "Verstehen and Critical Theory". There are two chapters on the positivistic critique of Verstehen and an introductory chapter on "The Classical Verstehen Position".

After surveying these debates, the author concludes that methodological pluralism is called for. We should make room for Verstehen approaches, yet without insisting that good explanations meet a particular Verstehen model. In his account, most of the Verstehen traditions "presume too narrow and restricted a view of the social sciences". (p. 1). He accepts the positivist claim that Verstehen is neither a method of verification nor a precondition for scientific understanding. At the same time he believes that the Verstehen method nonetheless can illuminate historical and economic events. An enhanced understanding of individual agent perceptions can help us discriminate amongst competing hypotheses, bring out previously undiscovered dimensions of a problem, or simply give us a direct intuitive sense of what is going on.

The final chapter defines "The Fundamental Maxim of Complex Methological Pluralism", which reads as follows: "Develop and use alternative methodologies and theoretical approaches that are relevant to the goals pursued and questions being asked!" (p. 250).

This conclusion, however sensible it may be, also lays bare the greatest fault of this book. It is boring.

I do not blame the author, since this work is less boring and clearer than its competitors. Nonetheless it convinces the reader that the entire topic of Verstehen, as presented, is a blind alley. Simple common sense could have brought us to methodological pluralism in the first place. So the book should more properly be regarded as a well-researched history of thought, rather than as a source of fruitful lessons for the future. It is a definitive treatment, but of a dead topic.

If we wanted to breathe life into the Verstehen topic, how might we proceed?

I would start with methodological pluralism and then ask how we might adjudicate between competing approaches. If we are allocating research effort at the margin, should we invest more resources into Verstehen approaches or non-Verstehen approaches?

This question, of course, cannot be answered apart from a particular institutional context, but I do believe a clear answer is present. The "hard" social sciences, such as economics, underinvest in the Verstehen method. Economists are too positivistic, relative to an optimum state of affairs. The soft social sciences, such as anthropology and sociology, overinvest in Verstehen. They would be better served by some more hard data, statistical rigor, and testable rational choice models, at their current margins.

In each case the overinvestment or underinvestment comes from pressures for scientific conformity. Fields, disciplines, and departments tend towards uniform methods, if only because that makes the work of others easier to evaluate. Because I share a common method with my fellow economists, I can judge their work with relative ease. I will vote to hire and tenure candidates of this kind. Every economist, acting in this manner, makes an individually rational decision. The collective result, however, is that the field ends up with too little methodological diversity, relative to an optimum. Knowledge generation would be better served by more diversity, but no individual institution wants to take on the costs of handling that diversity.

The central problem of methodology is thus an institutional one. We all know that methological pluralism is called for, but few of us have good ideas of how to get it in the proper doses.

One easy way of getting more methodological pluralism would be to abolish peer review. Amateur science is indeed marked by many different approaches. But the costs of this decision, in terms of quality, would be very high, so we must look elsewhere. We must consider what other rules of the game would get us more methodological diversity than we currently enjoy, and whether those alternatives would be worth the cost.

In short, methodology must incorporate more economic and institutional arguments if it is to make progress. The abstract philosophy of science only gets us so far. Once we end up in methodological pluralism, we do not know where to go next unless we make the analysis far more concrete.

Of course we face the paradoxical question of what method our analysis of methodology should use. Should we use Verstehen, positivism, or something else altogether? And if we use methodological pluralism, how should we invest our resources at the margin? This paradox suggests further that good method involves a bootstrap problem at all levels and that good method is rarely decided in advance by abstract arguments. For exactly the same reasons that the science of methodology is so important, writings in that tradition will rarely be effective in improving the practice of science.

Tyler Cowen
Department of Economics
George Mason University
Fairfax, VA 22030
Tcowen@gmu.edu

Pinkard's HEGEL: A BIOGRAPHY

PHILOSOPHY AND ECONOMICS: SEPARATE OR CONNECTED?

Paul Diesing

Cambridge: Cambridge University Press, 2000. Pp. xx, 765.

This book is a very detailed history of Hegel's life, based on voluminous letters and other historical sources. There have been several such biographies both recently and much earlier, and the ones I have read all differ. Pinkard describes all of Hegel's financial, career, interpersonal, and political troubles and how he dealt with them. But his main purpose is to show the development of Hegel's philosophic thought, by connecting it to Hegel's interaction with many other philosophers, and to his active responses to German and English political developments. The effect is to show Hegel's thought as a process, not a static systematic theory. The result is similar to Joan Robinson's account of Keynes' theoretical development in his weekly Saturday seminars from 1930–1936, except that Robinson ends with the completed, static *General Theory*. Pinkard continues through Hegel's whole life.

And, as with Keynes, Hegel's theoretical development didn't end at his death; it grew in several directions, and continues to grow (Pinkard, Epilogue). As Hegel said, "An old tree puts forth many branches". Pinkard's interpretation is one of those branches; one of his purposes is to show the errors of other branches. The first paragraph of the preface summarizes Hegel's thought; when I read it I was horrified, and felt I would be disagreeing with the whole book.

Research in the History of Economic Thought and Methodology, Volume 19A,
pages 245–247.
2001 by Elsevier Science B.V.
ISBN: 0-7623-0703-X

Then Pinkard went on: "Everything in the first paragraph is false". Oh, good. He was summarizing the other, wrong interpretations.

In Pinkard's interpretation Hegel's central political goal was fixed by about 1805: the task of us philosophers is to facilitate the development of the modern state that was begun in the French Revolution. We can do it by educating the professional class who will become the civil servants in various countries and then will gradually reform their government according to our (my) teachings of how the state ought to be. In his later years he developed his political views in more detail in response to European developments. He also developed his philosophy of nature and of religion more thoroughly.

Other Hegel interpreters, including some biographers, emphasize Hegel's empirical research as well as his arguments with other philosophers. They regard the political model described in his class lectures as a description of an already developed political system rather than a future ought-to-be. Of course his numerous discussions with other philosophers must have helped him interpret his empirical materials; but the actual state he was interpreting already existed. Hegel is quite explicit in the *Philosophy of Right* that one can only work out the structural essence of a society after that society has fully developed (Owl of Minerva). Presumably English society was the most fully developed modern society at the time, so its structure would be the model for later developing societies like France and Prussia. At any rate, his empirical studies were mainly of England. He read the English economists, especially Steuart and Smith, read an English daily newspaper and various journals, and also followed German and French developments. He came to see economic development, based in part on transport facilities, raw materials, and the Protestant ethic, as the main impetus for political development. Pinkard mentions the Protestant ethic and notes that Hegel read a daily newspaper, apparently for relaxation, but passes over the rest.

In this alternate interpretation Hegel was trying to work out the objective dialectic or underlying dynamics of the emerging modern society – the community-individualism dialectic, which Pinkard calls universalism-particularism, with individualism coming to dominate, the capital-labor dialectic, the self-destructive path dependence of law, the dialectic of marriage, the emerging and declining classes, and the interdependence of all these changing institutions. He also brought out the tensions, problems, and degenerative tendencies inherent in this structure. Pinkard passes quickly over the numerous institutional relations (pp. 473–489) though he had earlier emphasized Hegel's holism. He treats the *Philosophy of Right* as Hegel's rational construction of the sort of modern state that civil servants ought to establish. There is no mention of an objective dialectic; the structure is static.

Similarly, Pinkard interprets the *Phenomenology* as Hegel's response to various contemporary philosophers about the kind of philosophy of human consciousness that is appropriate to the modern world. An alternative approach would treat it as a history of the dialectical development of consciousness, ending with modern self-consciousness that contains the past within it. Pinkard agrees with this alternative approach and skims briefly over the long, complex history. However, he cannot explain the dialectic in detail; that would make the book much longer, and he had already explained the details in a 1994 book on the *Phenomenology*. For instance he passes over the master–slave contradiction in which independence produces dependence and vice versa; and he omits to mention how work, which produced the slave's independent consciousness, also induced the shift of consciousness to stoicism. Epictetus, the first stoic, was a slave.

So in each case the two alternative approaches are supplementary, but it would be impossible to work both out adequately in a short 666-page book. Indeed, Pinkard complains that he had to leave out many details of Hegel's relationships and concerns So one cannot fault him for leaving room for supplementary approaches; he had enough to do.

Garnett's WHAT DO ECONOMISTS KNOW?

ACADEMIC ECONOMISTS AND THE REST OF THE WORLD: MOSTLY POSTMODERNIST ESSAYS

Thomas Mayer

New York, Routledge, 2000.

This book is a mixed bag, both in terms of topics and methods, as well as quality – which varies from excellent to awful. It contains fourteen papers plus an introductory essay by the editor. These papers fall into four groups. Four papers center on the relation of economists to the public. Two deal with the related problem that tests of economic literacy do not look beyond neoclassical economics. Six focus on various themes that are united primarily by their postmodernist approach. The remaining two cover the relation of economists to foundations that sponsor economic research, and the philosophical foundations of econometrics. Most, but not all, of the papers have at least a postmodernist tinge. Garnett should not be criticized for such a lack of cohesion. The quality of papers is more important than is a unifying theme.

It is only fair at the start of this review to alert the reader to my biases. I am an insider critic of mainstream economics with a positivistic predilection, who accepts a moderate form of pluralism, but not postmodernism, except in its moderate form of challenging the pretension and arrogance of modernism. This makes me unsympathetic to many of these essays. Am I justified in criticizing

Research in the History of Economic Thought and Methodology, Volume 19A,
pages 249–259.
2001 by Elsevier Science B.V.
ISBN: 0-7623-0703-X

on the basis of my own paradigm essays that are at home in another? While not denying that paradigms are sometimes incommensurable I do deny that they always are. And even if paradigms are incommensurable in the sense that paradigm A solves problem X, while paradigm B solves problem Y, that does not mean that we cannot make the scientific judgment that, say A is superior to B – as long as we make no claim to apodictive knowledge. And a profession that accepts evidence at the 5% significance level can hardly be said to do that. It is therefore not clear what post-modernists mean when they reject the validity of criticism based on another paradigm. If they claim that it is useless, either because all paradigms are equally valuable, or because the choice between them is a matter of faith and not reason, then why waste time criticizing modernism? (Davis makes essentially the same point in his essay discussed below.)

I

Garnett introduces the book as partly a sequel to Deidre McCloskey's work on methodology that takes it a step further by presenting a postmodernist approach that refuses to privilege academic knowledge of economics over popular knowledge. This is, however, not an accurate description of the essays that follow: though it correctly describes some of them, others stress different, though sometimes related topics.

In the following essay, one that has a strong flavor of postmodernist moral outrage, Jack Amariglio and David Ruccio discuss the distaste that academic economists express towards popular economics, and how this is uncalled for, and inhibits effective teaching. Since such a theme runs through several of the essays I will discuss this paper in some detail. Amariglio and Ruccio argue that contrary to what academic economists think, the public's economics, which they call "ersatz economics", is not a set of unrelated, incoherent propositions, but has its own structure. It sees the economy as driven by the decisions of specific actors, such as the oil companies, and discussions of economics as driven by self-interest. That ersatz economics has a structure (though often only a rudimentary one) seems right. And that is important. If as teachers or as authors we want to change what the public thinks, it helps to know what it is currently thinking. I wish that this essay and subsequent ones had analyzed ersatz economics in much greater detail, particularly along the lines of Kahneman, Knetsch and Thaler (1986), a paper that Amariglio and Ruccio do not cite.

Amariglio and Ruccio give several reasons why economists might be so dismissive of ersatz economics: It reaches different conclusions that are sometimes vaguely left-wing. Moreover, acknowledging its legitimacy as a competitor would threaten academic economics Besides, surveys show that the public, particularly the lower classes, who might challenge the privileged status of economists, lack knowledge. This, they claim, feeds into economists' "fear of the masses". Thus, this essay, as well as Garnett's introduction and several of the other essays, conclude that academic economics does not deserve to be privileged over ersatz economics.

These authors are right in saying that academic economists should pay more attention to the economic beliefs of the general public, and that the public may have something to teach us. Here are some examples. Not so long ago while the public was condemning inflation, we economists mostly argued that it was not such a serious problem. Now we know better. In forecasting the inflation rate, households do not do all that badly compared to economists. In the late 1960s and 1970s monetary policy would have been better if the Fed had paid less attention to the advice of academic economists and more attention to the views of bankers. And more recently, would we really have been better off if the Fed had acted in accord with new classical theory as recommended by cutting-edge macroeconomists? We have a lot to be humble about. More generally if, as rational expectationists believe, the public knows, at least heuristically, the correct model, then its views deserve our attention. Furthermore, the public's thinking does not suffer from some biases that beset academic economics, such as a preference for results generated by complex rather than simple methods, and respect for the boundaries of academic disciplines. Academics do not operate entirely like truth-seeking missiles that, ignoring all distractions, inerrantly close in on the truth.

But let us not go too far. There are benefits to the division of labor. Economists devote more time and effort than do others to learning how the economy functions. In addition, while academic economists may perhaps be just as prone to self-interested pronouncements as are spokespeople for Phillip Morris or the Teamsters, most issues that economists discuss have few major implications for their self-interest. (And even when they do, self-interest is not always a good predictor. Academic economists were slow to come out against inflation even though academic salaries lag inflation) Their knowledge therefore *deserves* to be privileged in the same way as an auto mechanic's diagnosis of engine trouble deserves to be privileged over an economist's. To argue otherwise would require showing not just that economists have some biases, but that these biases are sufficiently important to offset the benefits derived from devoting greater effort to understanding the economy. And none of these essays in this volume do that. In true postmodernist fashion some seem to assume that once you have shown that unavoidable bias makes perfect truth unattainable, the extent of the error is irrelevant, a position that has the flavor of 19th century romanticism. Perhaps part of the problem lies in the use of the colorful word "privilege" that to modern ears cries out for the adjective "unfair". Substituting "preferred" for "privileged" might clarify matters.

II

Two of the other papers that relate the public's view of the economy to that of professional economists are solid empirical studies that are at home in a positivist framework at least as well as in a post-modernist one. In the first Arjo Klamer and Jennifer Meehan use the debate about NAFTA as a case study of the influence of economic analysis on public policy. They show that the proponents of NAFTA initially relied on the standard free-trade reasoning of economists – and got nowhere – in part because opponents of NAFTA disparaged the role of economists. But they did succeed when they relinquished economic analysis, and stated their case in a way that appealed to the public's concern about the character of the country and national pride and leadership. Klamer and Meeham's innovative and careful analysis is convincing for the case of NAFTA, but was that a typical situation or a special case? After all, NAFTA was not entirely an economic issue, foreign policy considerations were relevant. I suspect that this case was more typical than special, but only additional case studies – for which their essay shows the way – can tell. In addition, Klamer and Meehan criticize Paul Krugman for castigating politicians who ignore the advice of economists. They see this as an unwarranted privileging of academic economics, that among other things, ignores both the dissonance of the advice given by academic economists, and the differences in the rhetoric and metaphors of different speech communities.

In the next essay Robert Blendon et al. use public opinion polls to look at the public's assessment of economic conditions. When compared to what the data show, or to what economists believe, the public's assessment is generally much too pessimistic. The authors then suggest some reasons for this. But since their data come from surveys done in a single year, 1996, it is not clear how representative their results are for other years. For example, in time series data the public's inflation expectations (about which their paper does not provide any direct information) do not show much, if any, systemic bias towards pessimism. Again, further case studies are needed.

If these essays are right and the public thinks differently from economists, how should economists act when addressing the public? Richard McIntyre provides useful hints for an economist speaking to a labor-union audience. Since much of this would also be useful to economists addressing other audiences, it is a valuable contribution. His suggestions include knowing your audience, being partisan, but not too partisan, telling a compelling story, challenging the audience, but being careful not to intimidate it. (All this sounds obvious, but it should prevent the type of thing I once observed: an economist who in addressing an audience – much of which probably had taken at most Econ. 1 – presented a paper he had just published in the *Review of Economic Studies*, complete with differential equations.) Here, too, extensive further work could make a valuable contribution. In this essay McIntyre, who considers mainstream economics to be mainly apologetics and eschews the role of the disinterested scientist, also presents some of his strongly-expressed political views, in particular on the free-trade issue, where he grounds economists' advocacy of free trade in the special position of 19th century Britain.

III

In the first of the two papers on teaching economics Claire Sproul looks at the Advanced Placement Test and its underlying course curriculum that provides high school students with college credit. This course, she argues, is the last opportunity that many students will have to think about social issues. She shows that the microeconomics part of the exam relies solely on neoclassical theory without discussing its questionable assumptions, and that the macro-economic part relies on traditional macrotheory. She objects both because "teaching ideology-as-truth is toxic to democracy" (p. 218), and because such a control of secondary education by college curricula can stifle the creative possibilities open to high schools, for example discussing the role of class.

Sproul is right in complaining about the conformity and narrowness of vision that underlies the Econ. 1 curriculum. It is more difficult to know what to do about it. Yes, students should learn about Marxism and Austrian economics and institutionalism, not just neoclassical economics. But how much can be done in one short course? If these alternative approaches are presented just as brief glimpses they will not stick, and if they are presented in some detail there will not be enough time left for neoclassical economics. It may perhaps be better to cover the latter in some detail than to try to do too much, particularly if the reaction of students when presented with alternative viewpoints is likely to be: "if it is all controversial, why should I bother to learn any of it?". The root of the problem lies in something that Sproul mentions in passing, that this course may be the last time many students think seriously about social and economic issues.

A second and challenging essay on eduction by Grahame Thompson describes the introductory economics course at Britain's Open University. Instead of following Paul Samuelson's practice of stressing the unity of economics, this course stresses the diversity of economics by letting different "voices" speak (various parts of the course and its associated textbook were prepared by different teachers), and by teaching the limitations of the canon. Thus it considers ethical issues, racial and gender inequalities, and in discussing policy is sensitive to the losers as well as to the winners. It is also sensitive to institutional peculiarities, does not privilege perfect competition, and uses game theory extensively to put strategic interdependence in the foreground.

Such a course sounds exiting and much superior to the usual introductory course. But it raises several questions. Can the students absorb the messages of the extra voices in addition to the more conventional material, and if not how much of the latter has to be sacrificed? Are the students mature enough to accept the ambiguity of divergent voices? Such questions can only be answered by testing and talking to students who have taken the course. Moreover, as Thompson points out, the course was designed for a particular audience. Would other, presumably less mature students be able to absorb it? If the answers to these questions are favorable this course would go a long way towards curing what is now a serious deficiency in our teaching of economics.

IV

In his fine essay on postmodernism John Davis seeks to plot a future for postmodernism that tames its relativism and nihilism, so that it cannot be used by reactionaries just as readily as by progressives. (Davis deserves credit for being open about wanting to use postmodernism to provide a foundation for politically palatable conclusions.) He then follows McCloskey's suggestion of looking to the market as a model of how producers of different discourses interact within "multiple, evolving norms of conversation", norms which defy "abstract characterization". (p. 162), and are not amenable to the narrow mechanistic analysis of traditional economics. He illustrates how different discourses can communicate while maintaining their differences by the case of spousal abuse, and concludes by rejecting nihilism in favor of a "principled relativism" in which "particular discourses possess temporary and relative stability that enables their comparative investigation" though they cannot be arranged in a hierarchical order". (p, 166) Davis' paper is not easy reading, but is well worth the effort.

Suzanne Bergeron and Bruce Pietryowski deal with a novel that takes as its background an unusual development project in an African village, in which previously marginalized women hold most of the power, and the economic system is a mixture of capitalism, socialism and traditional economics. A considerable part of their essay is a commentary on the oppression of women and on the inferiority of conventional development theory to a theory that relies on cooperation, communalism and empowerment of women. More generally, they correctly point out that human activity as depicted in economic theory is only a part of human life, and that novels can remind us of the other parts.

The authors are right in saying that novels can provide us with knowledge. Does this mean that economists should read more novels? Not necessarily, because of its opportunity cost. Within the context of discovery (to use a rather unfashionable dichotomy) novels can certainly be useful by, as the authors point out, reminding us that economics deals only with a part of human life. (As Arthur Köstler once remarked, statistics don't bleed.) But novels have no role in the context of verification, because the novelist is allowed, indeed supposed to, make up his or her own "facts".

Two of the papers deal with hegemony. In one Ulla Grapard points out that one may express in humorous pictures what would be considered bad form if expressed explicitly in words. She therefore illustrates sexism in academia by analyzing pictures (cartoons and movie stills) that appeared in a popular money and banking textbook and disparaged women. Her illustrations come primarily from the 1977 edition. The 1993 and 1997 editions dropped most of the sexist pictures. But even these editions have two cartoons that she considers degrading to women. To me her analysis of these two cartoons, which in one case depends on the Freudian interpretation of the role of gold, seems a rather labored analysis of two good jokes.

But Grapard's demonstration of sexism in the earlier editions is compelling, and demonstrates well how such cultural attitudes infiltrate even seemingly harmless material. However, if read as more than a case study of a single book there are two problems. First how representative is the book? Second, does one need such a sophisticated analysis to show that sexism was rampant in our universities? We, or at least those of us who were in academia at the time, already know that at least into the 1960s sexism and racism pervaded academia. The current picture of academics as advocates of diversity and tolerance is one with a short history. What I did find surprising is that the sexist pictures survived at least until 1977 (Grapard does not tell us whether they were cut back before 1993), and that is a useful contribution.

Michael Bernstein's paper addresses an interesting question, why American economics achieved hegemony after World War II, and also the question of how

American economics helped American imperialism. How should one go about answering the former question? One might first ask how much of this hegemony can be explained by there being so many more academic economists in the United States than elsewhere, or by European academic economists placing relatively less weight on academic research and more on other professional activities (see Frey & Eichenberger, 1993). And one could look whether national differences in academic systems, such as the relatively free competition for eminent scholars among American universities, the small size of European academic markets, or the hierarchical organization of German universities, play a role. One could also take a historical turn, and look at the influence of refugee scholars in the 1930s, and at an earlier case of national hegemony in academic research, German universities in the late 19th. century.

Bernstein's paper proceeds very differently, relying heavily on post-modernist notions. Thus we are told that Harry Johnson's attempt to restructure along American lines the LSE's economics Ph.D.-program (which at that time had no graduate courses and consisted just of a thesis requirement on top of a B.A. program) was a "particularly graphic case of direct colonization" (p. 106) Others might read it as an attempt by Johnson (who was born in Canada and educated at McGill, Cambridge and Harvard) to modernize the British educational system. A significant part of the paper consists of describing the *imperialist* practice of American institutions to aid libraries that had been devastated by war or earthquake. (If that be imperialism let's have more of it.) The inclusion of an anti-Communist book in a list of books recommended by the American Library Association to devastated foreign libraries is said to make "clear" the "ideological content of the library effort" (p. 106). Bernstein also argues that WWII and the Cold War powerfully influenced American economics, citing input-output analysis, game theory, national income accounting, economic development, etc. Yes, it did, but if economists had refused to make their discipline helpful to an elected government couldn't they be blamed for defying the wishes of the people?

I am not sure that I understand Judith Mehta's vigorously postmodernist essay "Look at me, Look at you", To demonstrate the problem. here is the start of the main part of the paper:

> So what is in prospect? What will have to be done? Let's begin with a concrete idea rather than an abstraction: that what is being done exists in the event of its own production, an event in which we all participate. That is to say that here you participate with me in the search for a mode of seeing which will enable the unseen to become visible and a mode of re/presentation which will enable the unrepresentable to represent itself. It is a reflective act continually referring us back to the Subject. But who, or what, is the Subject? S/he is me, and s/he is you . . . (p. 38).

I *think* what this essentially says is that the message received in any communication depends in part on the recipient's preconceptions, but that we try to understand what the other is talking about. And in this case it is people like us.

Mehta goes on to tell us that economists are wrong to treat observations as objective facts, that the meaning of messages changes over time. "The act of capture immediately displaces and relocates the thing in a space which is out-of-its-time, but not timeless". (p. 41) She sees economists as using rational choice theory to constrain and discipline observations of human behavior. Authors ignore that they wear ideological blinders, and use the myth that observations are unaffected by the observer to avoid problems of human interaction. She grants that work done under scientific protocols is real, but claims that this is true also for other methods of inquiry, She discusses a Monet painting to show how it can be understood only by taking account of the means used to create it, and by relying on much sophisticated background knowledge. This she interprets as illustrating that the apparatus of rational choice theory and observations of behavior are interrelated, and then applies this insight to experimental economics. Finally, she argues against the suppression of ersatz economics by academic economics, using a game-theory experiment as an illustration of such oppression.

I am not persuaded. Yes – as we have known for a long time – facts are theory-driven, but that does not involve circularity unless the theory that drives them is also the theory we are testing. Moreover, even if the theory is confirmed by the "facts" only because the theory is generating its own "facts", if the theory is wrong, sooner or later independent facts will surface and cause trouble. Yes, all observation, classification and analyses distort, but if done well the distortion is small relative to the insight gained. Yes, we wear ideological blinders, but as the discovery of quantum theory demonstrates, they do not always blind us. Essentially, what Mehta seems to be saying is that absolute truth about the world is beyond human grasp. I agree – Hume established this more than two centuries ago – but I am perfectly satisfied if the results I obtain are significant at the 5% level, the coefficients are sufficiently large, and reverse causation, and multicollinearity, etc, are not serious problems. Why evaluate economics against a standard of pure truth that makes even dogmatic economists seem wishy-washy? Finally, do postmodernists, who claim to champion the common man against oppression by experts, not see the irony of expressing fairly straightforward ideas in an style opaque to the common man? And to others too.

In a brief, remarkably well-written paper Deidre McCloskey comments on the Amariglio-Ruccio and Mehta papers. Since she feels great affection for these authors she makes no pretence at objectivity. But she does present some mild criticisms of the Mehta paper for going too far, and defends herself from some criticism made by Amariglio and Ruccio. Using the telling terminology of "eighth floor" view (the view of the social scientist looking at people from a distance) and the "street floor" view of the people themselves, she argues that we can learn much from both. The economist's contribution is often to insist on accounting identities and the social point of view by exposing the self-interest inherent in many arguments by street floor denizens. But that, McCloskey argues, does not justify the vitriolic attacks of some academic economists on street-floor economics.

V

Radhika Balakrishnan and Caren Growth use their experience as program officers to elucidate how private foundations relate to economists seeking grants, and how they influence the direction of research. Since foundations stress policy issues, while graduate training deemphasises policy issues, there is a tension in the granter – grantee relationship. Other factors that reduce the efficiency of the grants system include faddishness and narrowness in peer reviewing, fiefdoms within foundations, the short-term orientation of foundations and their deemphasis of fundamental research, as well as academic fiefdoms and the tenure system's discouragement of innovative research by junior faculty. The authors devote separate sections to the National Science Foundation and to the following private foundations: Russel Sage, McArthur, Ford, Rockefeller, and Ohlin. Their essay provides useful information, but not as many practical hints for applicants as one might wish. Perhaps the authors might consider writing another paper or book on that. James Wible and Norman Sedgley discuss the philosophical foundations of econometrics, which they interpret as resting on critical realism. After discussing criticisms of econometrics by Keynes, Hendry, Summers, McCloskey and myself, the authors present their own version of critical realism that synthesizes ideas of Popper, Lakatos and Peirce. Its underlying themes are that the world has "many layers and levels of law-like order", the theory-laden character of observations, and the idea that "all science takes place in an economic context" (pp. 178–9). We thus need to juxtapose three theories, the one being tested, a second referring to how the data were constructed, and a third referring to the method (in this case econometrics) being used to relate the two. And we must do so within the limits of our budget constraint. Wible and Sedgley then apply this schema to econometrics to argue that patience and time for replication are needed before allowing econometric results to convincingly evaluate theory.

Moreover, inequalities in the financial resources that research programs have available may introduce imperfection in the competition among research programs. All this shows the precariousness of econometric work, and the long time that may be needed to for econometric results to confirm or disconfirm theories. This is a fine essay, but one that could do with much fleshing out. In particular, it would be worth seeing how this schema applies to actual instances in the history of economics and other sciences. But the authors cannot be blamed for not doing so in a short paper.

All in all, the strength of this book lies in the innovative questions it asks and in the quality of the answers given in some of the essays. Its weakness lies in the quality of the answers given in some of the other essays.

REFERENCES

Frey, B., & Eichenberger, R. (1993). American and European Economics and Economists. *Journal of Economic Perspectives*, 7, Fall, 185–194.

Kahneman, D., Knetch, J., & Thaler, R. (1986). Fairnes as a Constraint on Profit Seeking. *American Economic Review*, 76, September, 728–741.

Perlman and McCann's THE PILLARS OF ECONOMIC UNDERSTANDING: IDEAS AND TRADITIONS

PATRISTIC PERSUASION?

Glenn Hueckel

Ann Arbor: University of Michigan Press, 1998. Pp. xx, 639. ISBN 0-472-10907-3.

I. A BREATHTAKING SCOPE

Resting on a scholarship of awe-inspiring compass, this extraordinarily ambitious enterprise (the first of a two-volume study)[1] seeks to "develop a conceptual framework from which to organize the principal parts of the literature of economics" (p. xvii). But even so sweeping a proposal as this fails to convey the full scope of our authors' intentions, for they tell us that their organizing framework will encompass both "the *method of expression* of economic doctrines," as well as "the *substance* of those arguments" and, further, that it will reveal the influence in our literature of those "basic questions relating to the underlying nature of most modern 'western' social organizations" – namely, "such extraeconomic issues as Natural Law, and

Research in the History of Economic Thought and Methodology, Volume 19A,
pages 261–282.
2001 by Elsevier Science B.V.
ISBN: 0-7623-0703-X

natural rights, property rights, the design of social structures, and even philosophical questions regarding the basic elements of human nature" (p. xviii). In undertaking this monumental project, the authors aspire to join the company of Karl Pribram, Joseph Schumpeter, and W. C. Mitchell, to whom we owe what Perlman and McCann describe as the "three modern magisterial treatments of the history of economic thought, magisterial because they have created frameworks that serve to link together interpretations over time and across other boundaries" and because they are often found to be "uniquely interpretive" (p. 27). We are, in short, offered here the opportunity to join in an effort "to comprehend the contribution of economics to the overall social conversation," to "comprehend the connections between the current debate and its historical antecedents," and "to discover from whom economics has borrowed and why" (p. xx). The promise of so illuminating a rendering of our intellectual history cannot fail but to evoke from every serious reader well-deserved praise for the authors' breadth of vision. But such a response makes all the more painful the duty facing the reviewer who must report that the execution does not rise to the promise.

II. EXPLANATION AND PREDICTION

Unwelcome as they are, doubts regarding the efficacy of the authors' method emerge right from the beginning. The central element in that method – that which sets it "quite apart from the traditional historical approaches to the economics discipline" – is "a deeper appreciation of the influence of patristic thinking – that is, thinking along some paradigmatic lines determined by the cultural crucible in which the stuff of our minds is initially mixed." These "patristic legacies" are first described as no more (and no less) than the seminal origins of the discipline – "the nodes from which all of the ideas within the confines of the discipline radiate." But they soon seem to take on a troubling aura of determinism. We each have our own particular "patristic elements almost indelibly etched upon our minds," and these various ingrained legacies seem almost to determine the manner in which each scholar responds to the ideas inherited from the past. Thus, any claim to an "[u]nderstanding [of] the problems dealt with by economists ... begins with a grasp of the patristic legacies underlying their mind-sets" (pp. 2–3). Faced with forces apparently so deeply ingrained as to direct the path of scholarship, the reader at this point desires an explicit statement of those elements defining a "patristic legacy" or, at least, an enumeration of those legacies revealed in the development of economic doctrine. For this purpose, however, we must turn to the last chapter (where we find enumerated five classes of such legacies) or, better, to a recent article in which Perlman, in cooperation with a second co-author, seeks to restate the method elaborated in *Pillars* (Marietta & Perlman, 2000).

We find in that article that the passage of time since the composition of *Pillars* has done nothing to moderate the apparently deterministic tone by which the authors give expression to their vision of intellectual development. Indeed, it is revealed all the more boldly in their "central proposition," which is described as the principle "that the specific set of governing legacies that each individual economist possesses effectively guides his or her thinking" (2000, p. 151). Again we find that our "thought processes have been programmed," but now we learn that this occurs "without [our] knowing it." Nevertheless, "we seem to be somewhat intellectually path dependent" (2000, p. 155).

Of course, to observe that the ideas and rhetoric of the past influence the discourse of the present is to do no more than to state the very principle that animates all intellectual historians, and Perlman and his co-authors are no different. They propose to demonstrate "how economists of the past have consciously and unconsciously framed their arguments in light of their conscious and unconscious [use of] authorities." This knowledge of the "governing legacies" adopted by the authors of the past "leads not only to insight into the origin of their contributions, but also to how they framed their inquiries and what they excluded from them" (2000, p. 153). It can also indicate "how schools of thought more or less unite economists based on the degree to which members share similar authorities" (2000, p. 170). But for Perlman and company this principle is far more than a means to illuminate the past; it has apparently been transformed into a tool of prediction enabling its possessor to anticipate the responses of modern scholars and providing thereby a powerful means of persuasion. Thus we are assured that an "[u]nderstanding of opposing authorities . . . increases our ability to refute them and to persuade others based on a knowledge of their authorities, the fundamental basis of their system of thought." Such knowledge can influence our own thought as well since it may "increase our ability to be persuaded" by those "opposing authorities." In any event, the "crucial step" is a "conscious examination of our own authorities." Their influence on our work and its reception "can be counterbalanced by our conscious effort to understand the authorities of our audience and tailor our arguments accordingly" (2000, p. 185; see also *Pillars*, p. 578). Further, the power of prediction conveyed by a grasp of the "governing legacies" also alters our approach to the past. If we are to understand the methods of analysis adopted by past authors, we must "identify their central legacies," and to do that we must, of course, "consider the writings and biography of a particular author (or members of a school of thought)." But in view of the deterministic character of an author's "central legacies," intellectual biography apparently supplants intellectual history in this context. At any rate we are told that "[p]erhaps even more significant than the writings themselves are the crucial elements of biography, including the training and background, major influences in terms of teachers, readings, and experiences, and the intellectual milieu of a scholar's formative and productive years" (2000, pp. 170, 176).

By now the reader is all the more anxious for an unequivocal statement of those elements that define and delimit the notion of "patristic legacy." We come closest to that goal in Marietta and Perlman (2000, p. 154), where we learn that "the distinction between patristic legacies and mere (but important) ideas" rests on three tests. To rise to "patristic" status, an intellectual tradition must be "basic to a culture or significant line of thought;" it must "establish a framework for future thought," and it must "form the basis of a value system." But even this leaves a good deal of room for dispute in the application of these criteria, and the problem is made all the more complicated by the acknowledged characteristic of these legacies in operating below the level of consciousness. Although it is possible that authors can be fully cognizant of the legacies shaping their work, "perhaps more often the process is unconscious. Our intellectual baggage is difficult to deconstruct, but it constrains us whether we realize it or not" (2000, p. 154).

This obviously raises a question regarding the means to assent. By what means can Perlman and company persuade us that their "central proposition" is "an accurate conception of how scholars frame and understand economic problems" when the scholars themselves are unaware of the role played by their particular "patristic legacies" in shaping their thought? The only avenue open to us is to infer that influence from a reading of the scholars' work within the larger context of the discipline's intellectual development. But this requires that we bring to our reading of history a clearly delineated framework of identifiable "legacies" against which to judge the work of the authors who are the object of our study. Here too the only source from which we can construct that analytical framework is that array of intellectual traditions with which history presents us. Thus Marietta and Perlman tell us that, in determining "the larger set of frameworks . . . from which each scholar draws his or her personal subset," their "mannor of demonstration is a historical examination of the discipline, an examination not of ideas or theories, but of intellectual frames of inquiry that reappear consistently throughout the history of economics" (2000, p. 155). A reading of history, then, is central to the *Pillars* approach: it provides both the data to be studied and the organizational framework to guide the study. As Perlman and McCann tell us at the conclusion of the volume (p. 579), "We study the past because that is where the data are. More importantly, not only are the data there, but much of the data can be best explained by putting them into historical and cultural context," and " a major part of the cultural context is . . . one's patristic legacies."

The degree to which this approach proves fruitful will depend upon our ability to give that "larger set of frameworks" sharp definition and to keep it to manageable size. It is not enough that each of the organizational categories reveal a fundamental theme running through the work of several authors. The boundaries of each category must be fixed and clearly delimited as well, if we are to unambiguously distinguish each "legacy" from the rest. Categories defined so broadly or in terms that can be construed so elastically as to encompass a wide range of authors whose work is, at key points, highly disparate impede rather than enhance our understanding of the past. At the other extreme, if we find that our framework can encompass the relevant intellectual themes only by a proliferation of distinct categories, the whole enterprise risks descending to tautology, producing something rather like an intellectual history version of Ptolemaic spheres within spheres. Perlman and his co-authors are, of course, quite aware of these criteria; and they provide us, in the Marietta and Perlman article, with a complete enumeration of "the principal governing legacies in the evolution of economic thinking," together with the key names associated with each legacy (2000, pp. 156–170). In fact, the enumeration identifies twenty categories, all but two further divided into two or three "opposing concepts."

The reader who finds it daunting to be presented with an organizational structure composed of twenty major categories encompassing some forty distinct characteristics may find some consolation in the observation that just four of those categories bear the weight of the authors' argument in the *Pillars* volume. Furthermore, the structure adopted by our authors bears an unmistakable family resemblance to that found in Pribram (1983), one of those "magisterial" treatments that our authors take as their guides. The longest running of these four broad, recurring themes that shape the *Pillars* account is the continuing tension in what is described as "the first great 'debate' in economics: the Hobbes-Locke contest between communitarianism and individualism" (p. 221). The Hobbesian (communitarian) branch embraces all the usual suspects – the mercantilists and cameralists, the utopian socialists, Marx, the German historical school, and Joan Robinson – and, for added spice, includes the physiocrats and the later French writers through Cournot and Dupuit. But perhaps the most surprising addition to the group is Walras, whose work, we are told, is "permeated with communitarian concerns" (p. 572).[2] The alternative, Lockean tradition embraces nearly everyone else, chiefly the

British authors through Jevons and the Austrians. The two legacies are not, however, mutually exclusive. Thus, Adam Smith is said to have "embraced both traditions" (p. 572). Keynes, too, "variously stressed themes of individualism and communitarianism" (p. 396), and the American institutionalists managed to hold "a Hobbesian communitarian focus while absorbing along the way a strong commitment to a Lockean outcome, that is, the inherent nature of limited governmental power and largely the individual's right of choice" (pp. 572–573). The second of those four major themes is "utilitarianism," which in one or another of three distinct forms is said to link the work of all British authors from Smith to Keynes. Finally, we are presented with a methodological dichotomy between "Platonic idealism" or "Cartesian rationalism" on the one hand and, on the other, "Aristotelian inductivism" or Baconian observation and hypothesis (pp. 27–28, 573). From this contrast in method flows a congruent rhetorical divide between the mathematical formalism of the Cartesian branch and the "statistical reasoning" said to characterize the Baconian method. The Cartesian line is said to run from the physiocrats to Cournot and on to Walras. Baconian empiricism is associated with the political arithmeticians, and the British authors down to Marshall, though Jevons is said to have "embraced both legacies" (p. 574). A few additional "legacies" embellish the story – those associated with an appeal to natural law, the source of the right to property, and the role of a labor theory of value – but there are only four weight-bearing "pillars" in this structure.

Readers of the Marietta and Perlman article will find this observation curious in at least one respect. In the enumeration given there, the first of the twenty "legacies" cited is that concerning the definition of the "economic problem." This tradition is characterized as a division between the concepts of scarcity and uncertainty, both said to be "traceable to early interpretations of Genesis" as alternative views of "the punishment for the Fall of Man" (2000, pp. 160–161). This claim of a centuries-long dispute over the very definition of the discipline could reasonably be expected to play a leading role in any attempt to trace the development of our literature. Yet the question receives only the briefest treatment in the *Pillars* volume. The hare is started early on with the same appeal to contrasting interpretations of the Genesis story, but here we learn the "fathers" in whose work the two branches of this "patristic legacy" are said to have originated. Thomas Aquinas is said to have understood the punishment of mankind in terms of scarcity, while the work of Moses Maimonides, a twelfth-century physician, suggests the uncertainty view, from which we are to understand the "concept of uncertainty" as "an essential part of our patristic thinking" (pp. 18–19). Yet, apart from occasional, very brief passing mention, the concept does not resurface until the concluding chapter, where, in classic irony, it is offered as an example of the "quality of patristic legacies . . . to remain seemingly dormant for long periods of time." Only here do we learn that the concept had "lain dormant for several centuries" after Maimonides until it "surfaced again in the work of Cantillon" only to fall into "another period of dormancy" before it was rediscovered by Thünen and fully elaborated in the work of Frank Knight and in Keynes's *Treatise on Probability* (p. 570). The only substantive mention in the intervening five hundred pages is the observation that "Wicksteed's theory included a place for risk as a factor of production, a possibility mentioned earlier . . . by Thünen and even Cantillon, and expanded upon later . . . by Frank Knight and John Maynard Keynes" (p. 361). Curiously, Wicksteed's name is omitted from the list of adherents given in the Marietta and Perlman article. More curious yet is the omission from both the article and the *Pillars* volume of even the barest explanation of the role that the concept of uncertainty is supposed to have played in the work of Thünen and Cantillon. Indeed, Thünen receives nothing more than brief (unrelated) passing mention at two additional points in *Pillars*. Cantillon, however, commands the authors' attention for some six pages, but nothing there suggests that the notion of uncertainty plays any significant role in his analysis. The whole of that discussion treats his work as the "critical hinge" in the line running from the political arithmeticians to the national income accounting framework implicit in the physiocratic approach, adopting the Schumpeterian sequence of "Petty-Cantillon-Quesnay" (1954, p. 218). Although we find no mention of it in *Pillars*, it is true that Cantillon (1959, Pt. I, Chs, vii, viii, and xi) includes among his explanations of wage differentials the risk of premature death among skilled workers, but the bare notion that risk commands a differential return is so pervasive as to have little analytical import. Even Aquinas was willing to permit a profit to the partners of an enterprise as compensation for their risk (*Summa*, 2a, 2ae Q. 78, art. 2; 1975, p. 245). Apparently if we choose to trace our intellectual "legacies" back to this embryonic level, we may be unable to sustain the distinctions that Perlman and company wish to draw. But with no further explanation provided, the reader is simply left wondering what to make of the whole business.

III. A GUIDE TO THE PAST

Before turning to an evaluation of this "patristic" approach as a means to an understanding of the past, we must establish the appropriate standard of judgment. The issue arises because the appropriate definition of patristic categories and the allocation of authorities among those categories is conditioned by the use to which the resulting organizational structure is to be put – a point on which there seems to be some confusion among Perlman and his colleagues. In the Marietta and Perlman article, which gives the greater attention to the possibility of predicting the response of modern scholars and producing thereby a means of persuasion, we are told that any disagreement with their allocation of authorities among the various patristic categories reflects only "a tension" between "what the person's writings actually reveal" and the manner in which the author's work is commonly viewed. Their approach, we are assured, rests on the latter rather than the former (2000, p. 186, n. 7). One cannot object to this standard if the object is to construct a framework designed to predict the response of a modern audience. In that case, one would indeed be guided by the common interpretation accorded a particular authority even if that interpretation differs from that which would be produced by a close and careful study of the author's work. But as a means to an understanding of "what the person's writings actually reveal," such a strategy will only perpetuate the common error.

We find a related claim in *Pillars*, though that work aspires to a "magisterial" survey of the past. There too the authors plead for a less exacting standard on the ground that to "be magisterial is to give a strong interpretation; it is not to be authoritative." We are asked to consider their performance as we would those of "celebrated concert pianists" who succeeded in giving their audiences memorable interpretations of the composition though they "occasionally missed some of the notes" (p. 33). It would be churlish indeed to quibble over every missed note. Few mortals, least of all this reviewer, can claim an "authoritative" command over so broad a sweep of the intellectual landscape as that offered here. Nevertheless, the authors have set out to construct an interpretive framework "into which one fits the substance of past doctrines and dogma" fully aware that such "frameworks themselves then serve to shape perceptions further" (p. 570). There is more at stake here than a pleasant evening's diversion at the concert hall: errors in our interpretive framework produce errors in the vision of the past derived from that framework. When our goal is a deeper understanding of the composer's work, the performance must bear a close resemblance to the score as it left the composer's pen.

There can be no disputing the nearly superhuman character of the task that Perlman and his colleagues have set for themselves. The construction of an organizational framework that successfully reveals the key intellectual relationships and rhetorical forms spanning the whole of economic thought is a daunting enterprise, but like all human activity, this too is subject to defect. In this case, those potential flaws can themselves be formed into three classes. On the one hand, the categories chosen may be formed on characteristics that some authorities do not share. The ideas of those authors will then have to be pushed and pulled perhaps to the point of distortion to force a match where none exists. The alternative, as we saw earlier, is a scheme composed of categories defined so broadly, or so elastically, that they obscure key analytical distinctions between authorities. Finally, it is possible that our structure will permit a comfortable fit for all the relevant authorities but nevertheless obscure important alternative interpretations or alternative groupings revealing fruitful insights into the discipline's development. Readers will find cases of each of these types in the *Pillars* volume.

A. *"Misfits"*

Other readers may advance their own candidates for this dubious distinction, but many will likely agree that the author whose work suffers the greatest deformation as it is pressed into the *Pillars* structure is Adam Smith. Smith must fill two critical roles in that structure, both of which require that his work be endowed with characteristics that are quite foreign to it. He enters our story as "the first critical hinge in thinking," as scholars' attention turned from the issues raised by Hobbes and his critics to questions of a more strictly "economic" character (p. 36). Later, as the first of the British classical school – a designation that is defined by the theme of utilitarianism – Smith must serve "as an important figure in the rise of ethical Utilitarianism" (p. 230), a function not unlike that required of him in Pribram (1983, Chap. 9).

Upon his entrance, Smith is portrayed as one of those who sought to refute the Hobbesian view of man by which society is seen as " 'structured' to restrict the base nature of man" (p. 43). "Hobbes's man," we are told, "forms social bonds out of fear, that is, out of a desire to maximize his welfare by increasing his degree of safety" (p. 61). To draw the necessary contrast then, "Smith's man" is said to be "of a completely different mold," one that produces a being who "'has a natural love for society, and desires that the union of mankind should be preserved for its own sake, and though he himself was to derive no benefit from it'" (p. 59 quoting Smith, 1790, p. 88). The contrast is heightened all the more when we learn that "Smith's man is equipped with a very great moral sense, as well as a spirit of benevolence and even utility" (p. 61), the last quality preparing him for his additional role in shaping the "utilitarianism" of the British classicals. Finally, to drive the point home, the authors here dismiss as "rather an unfounded conclusion" Spiegel's (1983, p. 229) observation that "Smith rejects moral sense, benevolence, or utility as the basis of ethics." Readers familiar with Smith's *Theory of Moral Sentiments* will appreciate the irony in this, for Smith devotes Part VII, Section 3, Chapter 3 of that work to a thorough criticism of those theories that rest the "principle of approbation" in

a "moral sense;" Section 2, Chapter 3 of the same part is devoted to a criticism
of those systems that define virtue as benevolence, and Chapter 2 of Part IV is
devoted to a criticism of the view (which Smith attributes to Hume) that our
understanding of virtue derives from an appreciation of utility. Spiegel, was
right: none of these qualities provides Smith with the source for his system of
ethics; that distinction is reserved to his concept of sympathy.[3]

There is no denying that Smith's construction of sympathy conveys a degree
of sociability, but to insist that Smith understood us "to possess an innate need
to seek out human company and so form moral codes" is to dramatically
misconstrue his view (p. 64). To say, as Smith does (1790, pp. 13–14), that
"nothing pleases us more than to observe in other men a fellow-feeling with all
the emotions of our own breast" and consequently that "[s]ympathy . . .
enlivens joy and alleviates grief," is not to say that we have a "natural love for
society" and that therefore our actions are conditioned by an "idea of
community" (p. 59). On the contrary, the distance of the relationship influences
the degree to which we are able to enter into the passions of our fellows: "Men,
though naturally sympathetic, feel . . . little for another, with whom they have
no particular connexion, in comparison of what they feel for themselves"
(Smith, 1790, p. 86). It is true that we all are "in need of each others
assistance," but contrary to the *Pillars* claim (p. 61), nature does not rely on any
innate "feelings of community and belonging" to produce this necessary
association. Indeed, "[s]ociety may subsist among different men, as among
different merchants, from a sense of its utility, without any mutual love or
affection." All that is required is justice: "Justice . . . is the main pillar that
upholds the whole edifice." To ensure "the observation of justice, . . . Nature
has implanted in the human breast that consciousness of ill-desert, those terrors
of merited punishment which attend upon its violation, as the great safe-guards
of the association of mankind" (Smith, 1790, p. 86). As we pursue our own
interest, we avoid doing harm to those with whom we have no close connection
not so much from a "love for society" as from a fear of punishment. Indeed,
Smith's reference to that "love for society" quoted by Perlman and McCann
does not express his own view of human nature but is his characterization of the
opposing position that seeks to explain our desire to punish wrongdoers by
appeal to an appreciation of the utility of such punishment in preserving social
order. While Smith agrees that such an appreciation might enhance our natural
desire to enforce the rules of justice, he nevertheless insists that "it is not a
regard to the preservation of society which originally interests us in the
punishment of crimes committed against individuals." Smithian sympathy is
not the reflection of an "idea of community." It is rather an expression of the
capacity of individuals to bring home to themselves the emotion they imagine

would be produced if they were thrust into the situation in which they find another (see Smith, 1790, pp. 9–13). It is an emotional connection, however imperfect, between individuals by virtue of a common humanity:

> The concern which we take in the fortune and happiness of individuals does not, in common cases, arise from that which we take in the fortune and happiness of society. We are no more concerned for the destruction or loss of a single man, because this man is a member or part of society, and because we should be concerned for the destruction of society, than we are concerned for the loss of a single guinea, because this guinea is a part of a thousand guineas, and because we should be concerned for the loss of the whole sum. In neither case does our regard for the individuals arise from our regard for the multitude: but in both cases our regard for the multitude is compounded and made up of the particular regards which we feel for the different individuals of which it is composed.

Hence the concern which we feel for another "is no more than the general fellow-feeling which we have with every man merely because he is our fellow-creature" (Smith, 1790, pp. 89–90).

No point is served by a prolonged recounting of all the errors contained in the *Pillars* treatment of Smith's system, including, for example, the erroneous attribution of the virtue of self command to the exercise of reason (pp. 61–62) or the claim that Smith's system is a "rephrasing" of Shaftesbury's moral sense (p. 60). It is by now evident that readers familiar with Smith's ideas will be wholly unpersuaded by the argument offered in support of the claim that his *Theory of Moral Sentiments* reveals a "communitarian" outlook that contrasts with the more familiar "individualism" of *The Wealth of Nations* (see the designations accorded the two works in Marietta and Perlman, 2000, Table 1, p. 156).[4] Further the depiction of Smith as "an important figure in the rise" of utilitarianism, even if he is supposed to have "understood the problem with the consequentialism of Utilitarianism" (p. 239) will, in view of his repeated criticism of any attempt to form a utilitarian ethics, strike readers as quite extraordinary.

Smith is not the only author whose work must bend to the needs of the *Pillars* framework. Quesnay too would find that important nuances of this thought are sacrificed to that structure. As we have seen, one of the elements of the patristic approach is the principle that members of a "school" are united by an appeal to a common set of authorities. Quesnay, of course, is universally regarded as the founder of such a school. Hence, it must be possible to discern a common set of authorities uniting Quesnay and his physiocratic brethren. Here again, Perlman and McCann follow Pribram's lead (1983, Chap. 7) and depict physiocracy as taking the view "that economics could be considered an abstract and perhaps even deductive science along Cartesian lines" (p. 174). In keeping with their Cartesian disposition, the physiocrats are said to have exhibited a "lack of an empirical commitment" in that "[t]heir doctrines were not empirically verified" (p. 187). As to Quesnay himself, his views on these matters were apparently indistinguishable from those of the rest of the school. At any rate, we are told that the "epistemological position of Cartesian-Platonist rationality" exhibited by the *Encyclopédistes* "coincided with his own" (p. 177).

Readers familiar with Quesnay's *Encyclopédie* articles will certainly find it difficult to classify them under the Cartesian rubric. Long ago, Neill remarked on the strongly empirical character of those works, observing that "Fermiers" opens with the injunction to "consult the farmers themselves." Following his own advice, Quesnay produced a highly descriptive, comparative study of farming practices across the regions of France and England. The article on "Grains" reveals the like approach, "full of statistics and comparisons" (Neill, 1949, p. 549). Similarly, his earlier medical and philosophical works "reveal a mind more inclined to inductive than to deductive reasoning, one ready to check theory by observation and experiment, but a mind, nonetheless, which is not hostile to deductive, analytic reasoning when such a method seems appropriate to the subject under consideration" (Neill, 1948, p. 170). Hence, it is no mark of methodological discord that the same author who adopted the undeniably rationalist method of the *Tableau* could also produce the clearly inductivist "Fermiers." These catholic tastes in method make it "impossible to classify Quesnay under any of the convenient labels of 'rationalist,' 'empiricist' or the like," though Neill does insist that "Quesnay certainly was not a Cartesian" (1948, pp. 169–170). More recently, Vaggi (1987, p. 20) has abandoned the attempt to assign Quesnay to a single methodological tradition, opting instead to characterize his approach as "an original mixture of rationalism and empiricism." However, there is no mistaking the strictly rationalist stance taken by the other physiocratic writers, particularly Mercier and DuPont; but in this respect, "Quesnay stands apart from his disciples" (Neill, 1949, pp. 541–545; 1948, pp. 166–169). Here then we have a school founder whose epistemological attitudes clearly deviate from those of his disciples. Recognition of this distinction raises obvious questions regarding its source, the manner in which the school's adherents responded to their differences, and the consequences of those responses for the development and dissemination of their ideas. Yet none of these issues is made known to us by the *Pillars* framework.

Questions regarding the proper portrayal of Smith's system of ethics or of Quesnay's epistemology, though obviously important to a broad understanding of our intellectual development, cannot be said to fall within that central, defining core of issues that command the attention of all historians of the discipline. The treatment accorded Ricardo, however, will likely strike all readers as curious. To be sure, the authors' avowed intent to undertake here "an examination not of ideas or theories but of intellectual frames of inquiry," leaves little scope for a review of such familiar Ricardian themes as a labor theory of value, a falling rate of profit and stationary state, the principle of comparative advantage, the corn law, the "machinery question," and the like. After a brief passing observation that Ricardo's treatment of value "provided the impetus for the socialist challenges," we are promised a fuller discussion of the matter in the companion volume. We also learn that he "extolled the virtues of a free and unrestricted trade" but nothing of the analytical insight supporting that stance. Roughly half of the five pages devoted exclusively to Ricardo's place as "a pivotal figure in the history of economic thought" (pp. 263–267) are reserved for a discussion of his view regarding the right to property. This curious allocation reflects the authors' premise that the "definition and demarcation of the property right" is to be understood as a defining element of the "British Classical school" (p. 230). Ricardo's claim to membership in the school is apparently reflected in a conviction that the secure right to property is "the most fundamental of rights" – an attitude that is said to be "intimately tied to his view of the role of government, which he developed not only through his discussions with his acquaintances, but also through his . . . firsthand observations of the workings of government" while serving in Parliament (p. 264).

In support of these pronouncements we are offered a brief commentary on Ricardo's "Observations on Parliamentary Reform" (1952, pp. 495–503). However, we are not told that in assembling and editing Ricardo's works, Sraffa determined that this manuscript and a second on "Defence of the Plan of Voting by Ballot" were almost certainly composed at the suggestion of James Mill as an exercise in speech writing before Ricardo took his seat (Ricardo, 1952, pp. 489–494). While these papers certainly reflect Ricardo's views, they cannot be said to have been the product of his "personal experiences in the House of Commons" (p. 264). Nor are we told that the sentiments expressed in "Observations" regarding the character of the property right as "essential . . . to the cause of good government" are omitted from Ricardo's two Parliamentary speeches on Reform (1952, pp. 112–113, 283–289). Those occasions were devoted to Ricardo's call for the ballot, the subject of the second of those exercise papers, and a change which he viewed as central to any further reform: "unless the system of ballot were resorted to," he declared before the House in 1823, "it would be in vain to attempt any reform at all of parliament" (1952, p. 286). This is not to say that Ricardo changed his mind concerning the property right in the five years since composing the "Observations," only that the circumstances had changed. His earlier affirmation of the right to property was advanced to meet the fears raised by those who, in 1818, objected that an expanded franchise would "open the door to anarchy, for the bulk of the people are interested, or think they are so, in the equal division of property." As Sraffa observed, "The fears entertained in the disturbed years of 1818 and 1819 by 'those who have property to lose,' had by [the time of Ricardo's speeches] largely disappeared" (Ricardo, 1952, pp. 499, 491).

That Ricardo concerned himself with the cause of Parliamentary Reform is undeniable, but what all this has to do with Ricardo's "pivotal" role in the history of the discipline remains a mystery. Further, the reader is left wondering why, if the cause of Reform is critical to an understanding of that history, we learn nothing in the section devoted to Jeremy Bentham (pp. 243–251) of his earlier calls in the "Catechism of Parliamentary Reform" (1817) for the secret ballot, frequent elections, an expanded franchise, and uniform electoral districts. Neither do we learn of his influence in composing a series of resolutions on those points to be moved in the House of Commons in 1818, at the very moment when Ricardo, as an aspiring Member, was preparing his speech-writing exercise for Mill.

B. "Elastic" Categories

To be useful, a classification scheme must not only identify and group authors by those central elements of analysis held in common, it must also distinguish authors according to the substantive differences in the manner in which questions are framed and in the presumptions and logical structure underlying the analytical approaches directed at those questions. On this latter point, readers are likely to find the *Pillars* treatment of the "British Classical school" most bewildering. That designation is said to comprise an "'early' classical period," encompassing an intellectual line running from Hume through Smith, Bentham, Ricardo, Malthus, and the two Mills to Marx, and a later period defined by the work of Jevons, Edgeworth, Marshall, and Wicksteed (pp. 232, 308). This broad array is apparently held together by the principle of utilitarianism. At any rate, Perlman and McCann tell us that "[t]he focus of Utilitarianism as a shaper and organizer of classical British economic thought is thus a critical element in our interpretation" (p. 230). Indeed, "the ethical program that became known as Utilitarianism" is said to have formed "the basis for the emergence of British classical political economy" (p. 225). But this is a principle that can be found, in one form or another, throughout the history of economics – a discipline devoted to the study of utilitarian activity. As the authors remind us, Schumpeter (1954, p; 133) observed that even the scholastics "were utilitarian enough" in their analysis of market outcomes. We require some additional markers if we are to isolate the British classical authors within the broad stream of utilitarianism.

This is an unquestionably difficult task. Our authors quite properly caution us that the classical literature contains "distinct conflicts concerning the expression of the characteristic elements that serve to define the school," and they offer us a review of attempts made by other scholars to identify those defining elements. While those other efforts at definition overlap at a core composed of Malthus, Ricardo, and J. S. Mill at a minimum, they differ widely at the margins (pp. 228–230). Recognizing that any attempt to identify the limits of "classical" literature "will be somewhat subjective," Perlman and McCann propose a three-part test: "in conjunction with the definition and demarcation of the property right and the position of the individual vis-à-vis the community interest, the defining characteristic of the Classical school is a belief in the system as predicated on a Natural Law foundation" (p. 230). But this does nothing to narrow the focus: these are themes that run throughout the *Pillars* story from Plato at the beginning to Lionel Robbins at the end.

Neither do these themes isolate commonly-held analytical positions that can serve to set apart the classicals from other groups. We have already seen that the property theme does nothing to deepen our understanding of Ricardo's contribution. Further, the authors point to a "fundamental difference" on this issue between Smith (p. 67) and Hume (pp. 237–8) on the one hand and Bentham and the two Mills on the other. For the earlier authors, the property right pre-dates the institution of civil authority, finding its source in elements of human nature. Indeed, it is that "natural" appearance of a socially sanctioned right to property that, following Locke, calls forth the civil institutions designed to protect that right. For Bentham and the Mills, by contrast, the right to property is a creation of the civil law and therefore can be modified as circumstances require (pp. 246, 281, 293). A similar divide existed on the matter of natural law: Bentham "denounced the entire idea of Natural Law and natural rights as nothing more than fictions" (p. 247), while, of course, those concepts are central to Hume and Smith (see, for example, Haakonssen, 1989). Finally, on the contrast between the interests of the individual and those of the community, Bentham and James Mill are again the troublesome outliers; but here there seems to be some confusion as to the proper classification. We are assured that because the "individual is the fundamental unit of Bentham's society, . . . the needs of the individual in the provision of his personal welfare are granted primacy" (p. 244). Likewise, "The primacy of the individual was as central to [James] Mill as it had been to Bentham" (p. 262). But elsewhere we are told that in seeking "the 'greatest happiness of the greatest number,' . . . Bentham's Utilitarianism is fundamentally communitarian" (p. 249).

The problem of definition is further complicated by the requirement that the "classical" category include the "antagonists to classical political economy" – namely Marx, Robert Owen, and the Ricardian socialists (p. 232). The property right and communitarian themes serve as connecting threads here too; and they also reveal an interesting point of distinction between those who, like Owen, saw the property right as the creation of law and thus mutable, and Thomas Hodgskin, who apparently formed an attack on the profit return "based on the *legitimation* of the private property right" (p. 295). But now the classification scheme is modified to include additional defining themes. As to Marx, "his use of a moralistic rhetoric" is offered, in combination with his attention to the property right, as "evidence for his inheritance from the utilitarians;" and for both Marx and the Ricardian socialists, an "appropriation of the labor theory of value (evident in Locke, Smith, and Ricardo, among others) places [them] squarely within the classical tradition" (p. 298). Now, no one would object to the identification of a labor theory connection between Ricardo and Marx, but the extension of that thread to connect Locke and Smith as well will no doubt trouble readers familiar with Vaughn's reading of Locke (1980, p. 87) and those familiar with recent readings of Smith, any of the various editions of Blaug's text, for example (1997, pp. 48–52). Nevertheless, though it is extended too far for comfort, the labor theory thread does draw attention to an important element of analysis common to at least a portion of the so-called classical literature. In this respect, it seems a more fruitful frame of reference than a focus on attitudes toward property or on a possible communitarianism. Those latter themes provide neither a framework by which a clearly identifiable "classical" analysis can be distinguished from the rest of economic literature nor insight into the nature and structure of that analysis.

The situation is not much improved when we move forward to Jevons and the so-called "later classical" or "scientific utilitarian" literature. Here the points of distinction seem to be a move to an "ethically neutral" analytical structure and an emphasis on measurement. But again the reader is likely to come away from the *Pillars* treatment confused rather than enlightened. If limited to the price theory produced a generation or two after Jevons, there can be little objection to our authors' portrayal of that "later classical" literature as an "ethically neutral, or 'scientific' tool for decision making" (p. 308). But Perlman and McCann apparently mean the characterization to apply to Jevons as well. Much is made of the distinction drawn in the introduction to *Theory of Political Economy* between the higher order of "mental and moral feelings" on the one hand – obligations to family and friends or those felt by the statesman or soldier for the welfare of the nation – and "the lowest rank of feelings" on the other hand, those associated with the "mere physical pleasure or pain, necessarily arising from . . . bodily wants and susceptibilities" (Jevons, 1957, pp. 25–27). "It is here," we are told, "that Jevons's place in the history of economic thought is secured, for by dividing motives into higher and lower orders, and developing a procedure for handling quantitatively those lower values associated with the provision of base wants, he in effect removed the moral component from Utilitarianism" (p. 311). Now, this characterization will certainly come as a surprise to any reader who recalls the hedonist tone of Jevons's comment just two pages earlier that his theory "is entirely based on a calculus of pleasure and pain; and the object of Economics is to maximise happiness by purchasing pleasure . . . at the lowest cost of pain" (Jevons, 1957, p. 23; see Schumpeter, 1954, p. 1056). His attempt to narrow the scope of analysis to exclude the complications presented by the contemplation of benevolence or duty as motivating forces obviously did not preclude the application of his notion of utility to form normative judgments. Nor was it sufficient to avoid a rebuke from Marshall (1961, p. 101) that Jevons "led many of his readers into a confusion between the provinces of Hedonics and Economics."[5]

Even Perlman and McCann seem to be ambivalent concerning Jevons's claim to an "ethically neutral" system. Several pages after identifying for us the point where Jevons secures his place in history, they describe Jevons's contributions as "including the introduction of a methodology that reestablished Utilitarianism as a dominant economic and social (and even moral) philosophy" (p. 321). Later yet, we learn that the distinction between the "scientific Utilitarianism" of the late nineteenth century and the views of the earlier classicals is one of form rather than substance: "while the *essentials* of scientific Utilitarianism may not have been significantly different from those of its ethical heritage as exemplified in the works of Bentham et al. (throughout the later works one sees an emphasis on the same ethical problems dealt with by the Philosophic Radicals and their predecessors), the *form of the argument* – the rhetoric employed in the debate – was markedly distinct." Furthermore, the significance of this rhetorical shift is uncertain. At one point we are told that the changed rhetoric can be seen as "perhaps only marginally affecting the choice of the fundamental issues," though in doing so it certainly "influenced the conclusions derived." But just two sentences later, this change which "only marginally" affected "the choice of the fundamental issues" has now somehow "also redefined to a large extent the scope of the subject matter to which the later variants of the utilitarian doctrine were deemed relevant" (pp. 409–410). By now the reader's head is no doubt shaking in perplexity; and we shall leave off, pausing only to notice that the second of the proposed points of distinction noticed earlier – the desire for "a *measure* of utility"(p. 313) – strikes one as a peculiar choice in light of the criticism received by Jevons and his successors on that very point, objections that led later economists to abandon the notion of measurability after learning from Irving Fisher that an ordinal ranking is all that is needed, a story that is not told in *Pillars* (see Stigler, 1965, pp. 117–121, 152 and Schumpeter, 1954, pp. 1056–1066).

C. Issues Obscured

No organizing structure can capture all of the relevant relationships in a complex history. The test of such a structure cannot be its success in achieving an exhaustive classification but must instead turn on its ability to capture at least those filiations that are most helpful in deepening our understanding of that history. This is a necessarily subjective judgment, but it may nevertheless be useful to draw attention to some of those relationships omitted from the *Pillars* framework that seem to be most significant. We have already noticed the unsatisfactory outcome of the attempt to delimit the British "classical" and "neoclassical" literature by appeal to the theme of utilitarianism variously defined. After following Perlman and McCann in their pursuit of the several attitudes toward property and communitarianism, the reader will likely appreciate all the more the common description of the "classical" literature as emphasizing the process of economic growth in contradistinction to a "neoclassical" focus on efficient allocation within a static system (see, for example, Blaug, 1997, p. 278). Like all gross characterizations, this too is imperfect; but it does have the virtue of directing attention to clearly discernible analytical characteristics of the literature concerned. However, this means of distinguishing the two approaches does not arise in the *Pillars* framework.

Similarly, the Perlman-McCann resolve to express their organizing themes in terms of broad philosophical categories precludes a focus on the adjectival half of the "marginal utility" designation. Consequently, the famous triumvirate is not treated as a single unit. Each of the three authors is treated separately, embedded in his respective intellectual tradition: Jevons, as we have seen, as heir to British utilitarianism, Menger as writing in reaction to the German historical school, and Walras as exemplifying the "triumph of Cartesian rationalism." This structure obscures the long-standing fascination with the question of a so-called "marginal revolution" (see for example, Black, Coats & Goodwin, 1973). It is true that this separate treatment only makes the question all the more intriguing, but the question regarding the forces behind the independent appearance of the marginalist approach within three distinct intellectual traditions is never posed.

Furthermore, the decision to highlight the differences in each author's intellectual context obscures important elements of his analytical work. The portrayal of Menger, and "the Austrian perspective" in general, as "an antithesis to the German [historical] tradition," (p. 411) apparently leaves no room to acknowledge the long line of German scholars employing the ideas of diminishing marginal utility, the equimarginal principle, and even hints at marginal-product pricing in the factor market (see Streissler, 1990). We are told of Menger's dedication of his *Grundsätze* to Roscher, but we learn nothing of the array of those analytical elements fundamental to his theory that were available to him from the work of earlier, German scholars. Similarly, the portrayal of Walras as exemplifying a rhetorical shift "from a literary to a mathematical form of argumentation," said to secure his claim to the Cartesian tradition (pp. 465–466), obscures the substantive elements of his contribution. That contribution is summarized with the pronouncement that "Walras's achievements may be understood as formalism without development." "He produced," according to Perlman and McCann, "no grand theory of money or distribution or production" (p. 500). Is there then nothing of interest in, for example, his integration of money into his general system in Lessons 29–30? Are we also to dismiss his effort to solve the problem of disequilibrium trading in production by application of his *"tâtonnement"* process; or, more broadly, is the whole of Part IV of *Eléments* ("Theory of Production") to be ignored? Questions such as these are not likely to sit easily with many readers. Other readers will no doubt advance their own candidates for the role of "telling omission," but these few examples will suffice to illustrate the point.

IV. CONCLUSION

This is a book about fundamental principles. Indeed, it aspires to a "magisterial interpretation" of those principles. As one closes the cover on this attempt to employ our "patristic legacies" as a means to more fully understand the development of economic analysis, one is reminded of Schumpeter's warning: "It is the common failing of laymen, philosophers, and historians of thought to pay exaggerated respect to whatever presents itself as a fundamental principle." Impressive as it is in its scope, one fears that this work will only reinforce assent in Schumpeter's observation that "people do not always make use, in scientific work any more than in the practical concerns of life, of the fundamental principles to which they profess allegiance" (1954, p. 134). Furthermore, when, after the curtain has closed, one reflects on this performance, one cannot escape the impression that the artists missed too many notes. However, this must be taken as no more than a most provisional judgment. The volume before us tells only half the story. One can persist in the hope that the desired connections between the "patristic legacies" and the substantive development of economic analysis will be traced out in the promised second volume with each note given its proper weight. It is well for both reviewer and reader to recall in this context the sound scholastic principle that the "just price" of a commodity is to be determined by the "common estimation;" or, as Lessius stated the principle, it is "the price determined by peoples' general estimation" in the market. Such a rule is necessary because "private judgement is fallible," while the "[c]ommon judgement is less liable to error" (quoted by Gordon, 1975, p. 258). The same principle must apply in the realm of scholarship as well: the influence of the *Pillars* framework will be determined by the common judgment of scholars in the intellectual market place, not by the fallible, "private judgment" of a single reviewer.

NOTES

1. The second volume, *Pillars of Economic Understanding: Factors and Markets* (2000), is not considered here. Citations below indicating page numbers only refer to the volume under review. Citations combining the publication year of 2000 with a page number refer to the associated Marietta and Perlman (2000) article.

2. The authors have apparently had second thoughts regarding Walras's classification. In the Marietta and Perlman article, his name has been moved from the "communitarian" to the "individualism" line (2000, p. 156).

3. In assigning Smith to the utilitarian legacy, Perlman and McCann are again following Pribram's (1983, pp. 125–126) example, though his argument focuses more explicitly on Smith's sympathy, attempting to draw it into a utilitarian light. Others have advanced a similar reading, John Rawls among them, but such a view cannot be sustained. See Raphael (1975, pp. 96–97), Fitzgibbons (1995, pp. 45–57), and Griswold (1999, pp. 52–54, 139, n.23).

4. In an arresting *obiter dictum*, the authors advance another argument in support of their portrayal of Smith's *Moral Sentiments*, saying that he was "very clearly influenced by the physiocrats' communitarianism" (p. 572). This calls to mind the claim of those nineteenth-century scholars who found an "Adam Smith problem" in a perceived difference between *Moral Sentiments* and *Wealth of Nations* as to the presumed motivation for human action. Of course, it is now well understood that no such distinction can be sustained (see, for example, Griswold, 1999, Chap. 7, or Young, 1997). Nevertheless, it used to be said that Smith was led to abandon the principle of sympathy developed in his earlier book in favor of a reliance on self interest in the later work after he breathed the materialist atmosphere of French philosophy during his visit to the Continent (Oncken, 1897, pp. 444–445). However, this argument works no better for Perlman and McCann than it did for the proponents of that earlier Adam Smith problem. The 1759 publication of *Moral Sentiments*, which is supposed to contain his pretended "communitarianism," clearly predates his encounter with the physiocrats during his stay in France from 1764 to 1766 (on that visit, see Ross, 1995, Ch. 13).

5. It is true, as Perlman and McCann observe (p. 331), that Marshall took notice of this Jevonsian hierarchy of motives, but for him, "no new difficulty is introduced by the fact that some of the motives of which we have to take account belong to man's higher nature, and others to his lower" (1961, p. 16).

REFERENCES

Aquinas, T. (1975). *Summa Theologiae*, vol. 38: *Injustice*. Trans, Marcus Lefébure, O. P. London: Eyre & Spottiswoode.

Bentham, J. (1817). Catechism of Parliamentary Reform. In: J. Bowring (Ed.), *Works*, vol. 3. Edinburgh: William Tait, 1843.

Black, R. D. C., Coats, A. W., & Goodwin, C. D. (Eds) (1973). *The Marginal Revolution in Economics: Interpretation and Evaluation*. Durham, North Carolina: Duke Univ. Press.

Blaug, M. (1997). *Economic Theory in Retrospect*, 5th ed. Cambridge: Cambridge Univ. Press.

Cantillon, R. (1959). *Essai sur la Nature du Commerce en Général*. Trans, Henry Higgs. London: Frank Cass.

Fitzgibbons, A. (1995). *Adam Smith's System of Liberty, Wealth, and Virtue: the Moral and Political Foundations of The Wealth of Nations*. Oxford: Clarendon Press.

Gordon, B. (1975). *Economic Analysis before Adam Smith: Hesiod to Lessius*. New York: Macmillan.

Griswold, C. L., Jr. (1999). *Adam Smith and the Virtues of Enlightenment*. Cambridge: Cambridge Univ. Press.

Haakonssen, K. (1989). *The Science of a Legislator: the Natural Jurisprudence of David Hume and Adam Smith*. Cambridge: Cambridge Univ. Press.

Jevons, W. S. (1957). *The Theory of Political Economy* (5th ed.). New York: Augustus M. Kelley, reprint, 1965.

Marietta, M., & Perlman, M. (2000). The Uses of Authority in Economics: Shared Intellectual Frameworks as the Foundation of Personal Persuasion. *The American Journal of Economics and Sociology, 59*, 151–189.

Marshall, A. (1961). *Principles of Economics* (9th (*variorum*) ed.). Ed. by C. W. Guillebaud. London: Macmillan.

Neill, T. P. (1948). Quesnay and Physiocracy. *Journal of the History of Ideas, 9*, 153–173.

Neill, T. P. (1949). The Physiocrats' Concept of Economics. *Quarterly Journal of Economics, 63*, 532–553.

Oncken, A. (1897). The Consistency of Adam Smith. *The Economic Journal, 7*. 443–450.

Pribram, K. (1983). *A History of Economic Reasoning*. Baltimore, Maryland: Johns Hopkins Univ. Press.

Raphael, D. D. (1975). The Impartial Spectator. In: A. S. Skinner & T. Wilson (Eds), *Essays on Adam Smith*. Oxford: Clarendon Press.

Ricardo, D. (1952). *The Works and Correspondence of David Ricardo*. Ed. by Piero Sraffa. Vol. 5, *Speeches and Evidence*. Cambridge: Cambridge Univ. Press.

Ross, I. S. (1995). *The Life of Adam Smith*. Oxford: Clarendon Press.

Schumpeter, J. A. (1954). *History of Economic Analysis*. New York: Oxford Univ. Press.

Smith, A. (1790). *The Theory of Moral Sentiments* (3rd ed.). Ed. by D. D. Raphael and A. L. Macfie. Oxford: Oxford Univ. Press, 1976.

Spiegel, H. W. (1983). *The Growth of Economic Thought*, rev. and exp. ed. Durham, North Carolina: Duke Univ. Press.

Stigler, G. J. (1965). The Development of Utility Theory. In: *Essays in the History of Economics*. Chicago: University of Chicago Press.

Streissler, E. W. (1990). The Influence of German Economics on the Work of Menger and Marshall. In: B. J. Caldwell (Ed.), *Carl Menger and His Legacy in Economics*. Durham North Carolina: Duke Univ. Press.

Vaggi, G. (1987). *The Economics of François Quesnay*. Durham, North Carolina: Duke Univ. Press.

Vaughn, K. I. (1980). *John Locke: Economist and Social Scientist*. Chicago: University of Chicago Press.

Young, J. T. (1997). *Economics as a Moral Science: the Political Economy of Adam Smith*. Cheltenham, U.K.: Edward Elgar.

Henderson and Davis's THE LIFE AND THOUGHT OF DAVID RICARDO

Samuel Hollander (Edited by Warren J. Samuels and Gilbert B. Davis)

Kluwer Academic Publishers, Boston/Dordrecht/London, 1997, pp. xii–679.

I

The outside cover of this work states the authors to be John P. Henderson and John B. Davis; but the title page has John P. Henderson "With Supplemental Chapters by John B. Davis," and the Introduction – by one of the two editors (Warren J. Samuels) – is devoted entirely to Henderson who died in February 1995, after over a decade of debilitating ill-health. Here it is explained that Henderson's manuscript, as it had been left by him in 1983, contained only the first eight of the thirteen chapters that comprise the published version: "The Multiple Role of the Biographer;" "The Sephardic Heritage in English Society;" "The Family Heritage: Eighteenth-Century Finance;" "Boyhood in London and Amsterdam;" "The Taming of Tradition;" "The Gestation of an Economist: Early Financial Career;" "Malthus and the Corn Law: Ricardo and his Circle;" and "Ricardo's *Principles* and the Question of Value." Chapter IX,

Research in the History of Economic Thought and Methodology, Volume 19A, pages 283–295.
2001 by Elsevier Science B.V.
ISBN: 0-7623-0703-X

entitled "Friendly Critics: Malthus and Ricardo on Political Economy," is constructed from two pieces which Henderson had indicated were to have been the basis of the chapter; and Chapter XIII, "A Critique of the Twentieth Century Perspective," was selected by the editors as a Conclusion from materials in a paper "prepared by Henderson for a meeting, apparently in Texas," which from internal evidence took place in 1977 (see editorial comments, pp. 1, 615, 626). There is considerable overlap between the last and the first chapters. The three remaining chapters – X: "A New Career in Politics;" XI: "Equivocation: the Effects of Machinery on the Demand for Labor;" and XII: "The Search for a Measure of Absolute Value" – are by John B. Davis, a former student of Henderson who (Samuels informs us) "shared many of Henderson's views about Ricardo" (p. 10). I shall, however, limit my observations to the work of Professor Henderson.

To review a book based largely on a very incomplete manuscript dating back at least 17 years, and even that part unrevised by its author, is a difficult assignment in itself; the fact that its author is deceased ought not perhaps be a consideration but one tends to be more sensitive in such a case especially if his poor state of health for many years precluded the revisions and corrections that he may well have realized were required. All in all, the reviewer's responsibility to be fair and judicious are greater than usual in the present instance.

II

Professor Henderson states that his "primary purpose" is "to tell the life story of David Ricardo," and only – as he himself put it – *secondly* "to present my perception of his economics and what it meant" (p. 17). These two objectives, however, merge shortly thereafter, with Ricardo the *economist* taking center stage:

> Although this book must of necessity be mainly concerned with Ricardo the economic theorist, it is at the same time an intellectual biography, not merely an exposition and analysis of Ricardian economics. Of course, much attention will be given to the development of his economic ideas, but an intellectual biography is also an integrative narrative of the personal, social, and theoretical aspects of a particular individual's life. Accordingly, it is essential to appreciate and understand not only Ricardo's economics and his views on sociopolitical matters, but also what manner of man he was and how he reacted to his environment, that broad social milieu which encompasses not only friends, allies, family, and comrades in arms, but also antagonists (p. 21).
>
> An intellectual biography draws from several disciplines. Of first importance is the development of the individual's intellectual powers, and this would suggest a biographer must have first-hand knowledge of the field in which his subject was interested. That is, because David Ricardo was a political economist, someone familiar with economics should

be his biographer. But the biographer must also be in sympathy with his subject's theoretical orientation, if for no other reason than to be able to present a perceptive image of the individual's contribution to the field . . . (p. 42).

The first four chapter headings cited above convey something of the background painted by Henderson – the focus of attention is throughout to be on the significance in "understanding" Ricardo of his early traditional Sephardic upbringing. What must be clarified is whether Henderson went so far as to claim that *Ricardo's economics* was itself a product of that environment, and if so in what way. For it is a fact of life that he was raised in a traditional Jewish – specifically Sephardic – household. Any "Life of Ricardo" must include the story of his upbringing and education and if it were simply a story that Henderson intended to relate he cannot be faulted – apart from certain errors of fact that he might have discovered under happier circumstances. If, however, Henderson intended to relate Ricardo's *economics* to his early environment the matter becomes far more complex and controversial. To decide whether or not Henderson succeeded in his endeavor depends on being absolutely certain of what that endeavor was.

Consider then some pointers: "Of great importance in the life of David Ricardo was the fact that he was born a Jew in English society. More important, he was reared in the Sephardic community of London" (p. 76). Henderson proceeds to relate features attributed to Ricardo's "character" to that upbringing:

> The insecurity of the London Sephardic enclave in the seventeenth and eighteenth centuries led to the development of a system of self-regulation and sanctions. Out of this system there emerged a strict moral code which imposed constraints upon members of the stock exchange, for the bourse was the economic center of the community. Personal accountability and a strong sense of responsibility to the group whether in the synagogue or in the stock exchange, was not only expected but also demanded. It was within this atmosphere that David Ricardo's character was molded, the product of a social system responding to the prejudices of English society (pp. 76–77).

I am not competent to debate the nurture-nature balance; perhaps Professor Henderson is right. But the immediately following remark is certainly open to question: "That system within the Sephardic enclave also fostered egalitarianism, responsibility and tolerance" (p. 77). This perspective is strongly insisted on:

> The origins of Ricardo's political views, and his strong egalitarian and humanitarian instincts, have typically been associated with his friends, James Mill and Jeremy Bentham, both ardent philosophical radicals in the tradition of the eighteenth and nineteenth centuries. But as subsequent analysis will reveal, Ricardo's ideas predated his association with his radical friends and should more correctly be traced to his Sephardic origins. Certainly, he was reinforced in his social outlook through his friendship with James Mill,

and to a lesser extent Bentham, but the democratic spirit Ricardo championed originated
not in Mill's Scotland, but in the Sephardic enclave of London (p. 24).

As for "tolerance," I fear that Professor Henderson was *too* tolerant. The
evidence at our disposal points to a dictatorial, rigid and outmoded Synagogue
leadership (exercised by the so-called "Mahamad") at Bevis Marks –
duplicating that in Amsterdam – that may well have contributed to the defection
of numerous celebrated families (apart from David Ricardo and several of his
siblings), including the families Basevi, D'Israeli, Samuda, Uzzielli, Lopez and
Ximene (see Gaster, 1901). David's defection can, I submit, be better
understood if account is taken of this general phenomenon. At one point
Henderson in fact alludes to the dictatorial character of the administration,
though somewhat apologetically (citing Hyamson): "The Mahamad had
considerable power over the Yehidim, and they wielded this power sometimes
somewhat dictatorially, but their object was the welfare of the Community as
a whole, and this was recognized" (p. 76; emphasis added). And there is also
a reference to "the shadow of the Synagogue, the reminder of tradition,
conformity, and purpose" (p. 163; my emphasis).

Nor do I see much evidence of "egalitarianism." We are provided with
photographs comparing the London Burial Ground of the Sephardim and that
of the Ashkenazim to demonstrate that the former made no distinctions among
the deceased with respect to graves – a point written into the constitution of the
community[1] – unlike the latter: "The great diversity in the grave monuments in
theAshkenazi cemetary is testimony to the fact that status was tolerated and
encouraged" (p. 52).[2] The problem I face arises from a photograph I have
before me of the Spanish and Portuguese cemetary on Mile End Road – the
same as in Henderson's photograph – that does not show a section with
gravestones level to the ground, but rather one containing an array of finely
differentiated raised tombs (*The Jewish Encyclopaedia*, VIII, p. 158). An
engraving by Ruysdael of some truly splendid tombs in the cemetery of the
Sephardic community in Amsterdam is similarly problematic for Henderson's
case (see *The Jewish Encyclopedia*, I, p. 544). Apart from this, the hostility to
intermarriage with Ashkenazim – relatively disadvantaged newcomers to
London – amongst many in the Sephardic community well into the nineteenth
century is easily documented and points away from tolerant egalitarianism.[3] All
this matters greatly for the interpretation offered in this book of Ricardo's
"view of political economy and politics" (p. 77).

But Henderson's environmental theme extends much further, in that he
ascribes Ricardo's economic *analysis* to his origins:

> . . . his view of political economy and politics and his ability to analyze the operation of the
> English economy in no small measure were attributable to the system of values and

practices of the enclave of which he was a member. His environment and parentage helped to make him a stockbroker and a financier, and his origins in a community apart gave him the objectivity of an outsider *that made his economic analysis distinctly atypical among English economists* (p. 77; my emphasis).[4]

And if we turn to the concluding chapter, we find this line fully confirmed, indeed taken one step further by reference to Sraffa and adoption of what has since come to be known as neo-Ricardian historiography:

> Like Ricardo before him, Sraffa early in his career rejected the approach of what is best described as the Marshallian "scissors" analysis, with the result that demand never became an active participant in either of their schemas. Moreover, Sraffa's own particular orientation preceded his editing of Ricardo's *Works*, and perhaps it was because of this orientation that he was capable of solving many of the so-called Ricardian "muddles." Neither Ricardo nor Sraffa, for example, were products of the English public school system, and even though Ricardo lived in England all of his life, it was his non-English approach to matters that made his economics so controversial, not only during his own lifetime but long after.
>
> Ricardo's non-English approach to economic matters, his highly theoretical instincts about the pressing economic issues of his time were grounded, as has been argued in earlier chapters, in the emphasis upon logic and deduction which were a part of his Jewish tradition. That Talmudic logic frequently is tortuous cannot be denied, such as the rule that the banning of marriage between holy festival days can only be violated by a man who takes back his divorced wife. Nevertheless, Talmud studies fall under the general rubric of the study of logical processes. It is better pedagogy to use the Talmud, for example, than the Iliad and the Odyssey in order to attempt to train a logician. The classics, Greek literature, and the history of the English monarchy, each was foreign to Ricardo, and in the grand tradition of a logician, his most frequent openings were: "let us assume," or "let us suppose." In a sense, Ricardo needed a non-Englishman to understand his economics, and for this reason Sraffa was an ideal choice (pp. 637–638).

I conclude that it is fair to attribute the strong view to Henderson that Ricardo's very method in economics reflected his Sephardic origins in general and Talmudic training in particular.

III

I shall return to examine Henderson's position on the foregoing issue in Section VI. But first let us focus on the simpler issue, the "story" of Ricardo's Jewish background. We have already touched on aspects of Henderson's account of the Bevis Marks Synagogue. But there is more. I came away feeling rather like a native in darkest Africa might feel reading an account by an anthropologist-explorer who did not always get things *quite* right. For example: the reading of the Torah in the Synagogue requires ten not seven males over thirteen years of age; the "symbol of the synagogue" is not the *bimah* or reading stand – if there is a "symbol" it is the ark containing the Torah scrolls; and there *is* a priesthood (see on these matters, p. 63). It is by no means certain – as is implied on p. 145 – that Ricardo was taught Hebrew "as a living language", or that his "earliest experiences would have been the daily ritual of his father donning "the tifillion (sic!) and prayer shawl for morning prayers at Bevis Marks" (p. 145); only if his father said his daily prayers at home would David have observed him laying *Tefillin* (phylacteries), unless the young boy also went to the Synagogue during the week, and this we do not know. The *"kidush"* is recited on Friday night – and a short version on Saturday – not to "reconfirm that the Jews are God's chosen people and that they have a responsibility for their convenant" (p. 145), but preeminently to recall that God ceased all creative work on the sixth day. *Brit mila* (circumcision) is in fact obligatory on the Sabbath for healthy, eight-day old infant boys, and there is certainly "conviviality" on the Sabbath (p. 145). On the other hand, one positively does not recite the *kidush* for a deceased person (p. 166), for that rite is reserved for joyful occasions (Sabbath and festivals) – we are not talking about an Irish wake; rather it is the *kadish* that is recited. If the youngest son were to ask: "Mali nish ta moh"? at the Passover Seder (p. 146), he would be in big trouble, especially if his father had been financing his Hebrew education – doubtless "Ma nishtana ... "was intended or a printer's error occurred in this case. Rosh Hashana is a two-day not a seven-day holiday (p. 146). And the Sabbath Torah portion is divided between seven not six members (p. 154). A far more serious error is the affirmation that "[t]he Babylonian Talmud was followed by the division of the Jews who eventually became known as the Sephardim, while the Palestinian Talmud was adopted by the Ashkenazim" (p. 62). Where did Henderson obtain this *canard*? The Ashkenazi academies devote themselves almost entirely to the Babylonian Talmud.

It is regrettable that these slips were not caught before publication, since they cumulatively may have the effect of undermining confidence in Henderson's account. And this would be sad indeed, since there is much that is excellent on Ricardo's life and times as I shall now indicate.

IV

Henderson, to my mind, is best on the financial and political history of the eighteenth and early nineteenth centuries (pp. 86f., 195f., 241f., 256f., 354f.) the social history of London (p. 121f), contemporary publishing and reviewing practice (pp. 220f., 449f., 459–460, 477f, 491f.), Ricardo's own financial affairs (pp. 212f., and Chapter VIII *passim*), his various "careers," and his personal relationships with other economists (throughout, but especially Chapters VII and VIII). I find all this, including the numerous digressions on the lives and careers of Ricardo's friends and colleagues (particularly Malthus and James Mill), refreshing, entertaining and above all instructive.[5] Much of the data can be found in Sraffa's edition of the *Works and Correspondence* – Henderson's primary source.[6] But the constructive distillation is to Henderson's credit, and I came away from the book with an enhanced sense of the period and the actors. Furthermore, the presentation is lively, conveying the impression that Henderson thoroughly enjoyed what he was doing.

V

What now of the economics narrowly defined? As we have indicated, Henderson subscribed to the Sraffa – actually the Marx-Sraffa – view of Ricardian economics and its role in the development of the history of economic thought. The closing passages of the book – following a discussion of the "real value" issue – elaborates thus:

> But philosophically, Ricardo and Marx struggled with one identical problem, namely that the distribution between wages and profits was independent and prior to the determination of the prices of commodities as these circulated throughout the system. On this score, Ricardo was concerned with the differences between the "real value of commodities" when considered as a whole, and the prices which the market assigned them individually. This was the sense in which the inclusion of durable capital required the development of the labor theory of value so far as the theory of exchange was concerned. As an explanation of aggregate profit, the labor theory of value appeared quite adequate, but as an explanation for the process of actual exchange, it required further elucidation to show just how prices followed from the values determined at the time of production (p. 655).
>
> Ricardo's Ricardo was the economist's economist. Far from being muddled, he possessed insights into the intricacies of economic theory which escaped not only his contemporaries, but the traditional followers of orthodoxy. Only the refugee bookworm in the British Museum appreciated and understood the theoretical problems with which Ricardo struggled, despite the fact that the bookworm was unaware of Ricardo's last attempts to resolve the conundrum. A second refugee provided most of the missing pieces, even though there remains some doubt as to whether Sraffa's *Production of Commodities by Means of Commodities* has resolved all of the problems of the transition from values to prices (p. 656).

There is too the forceful rejection of Marshall's reading of Ricardo; the Marxian ideological interpretation of the marginal utility developments, in which context Henderson draws on Ronald Meek and Maurice Dobb; and the notion of a "decline" in Ricardo's influence after the 1830s (pp. 526, 621f.),[7] the latter, of course, an important feature of Marx's position.[8]

I am not in sympathy with this perspective on Ricardian economics and its place in the development of economic thought, and see considerable merit in the Marshallian view. But I do not intend to argue the case here, merely to state where Henderson's orientation lies. I must though point out that Henderson, who puts great weight on Torrens in accounting for the alleged decline in Ricardo's reputation (pp. 543–546), must have been unaware of Torrens' candid retraction of his original objections to Ricardian theory – "Some of the commentators on the doctrines of Ricardo appear to have fallen into the misconception, that, in altering his nomenclature and in modifying his principles as varying circumstances required, they refuted his theory of profits" – and that he came to defend the inverse wage-profit relation and Ricardo's proportions-measuring money against the objections of Senior and Jones, in a belated about-face that he himself attributed to the influence of Longfield (Torrens 1844, pp. xxiif, xxxvi, li–lii).

VI

I return now to the central issue: the author's linkage of Ricardian theory to personal experience and general environment. I shall raise two problems, first the connection made between Ricardo's method (as Henderson describes it) and his alleged Talmudic training in his early years (above, p. 285). I say "alleged" precisely because Henderson does not make it clear whether (or to what extent) Ricardo received Talmudic training – he merely takes it for granted. Now Henderson maintains – against Heertje (1975, p. 79) – that Ricardo attended the Amsterdam Talmud Torah from 1783 to 1785 rather than a private school. Let us accept for argument's sake that this was so. The Talmud Torah was a *preparatory school* for boys from five to thirteen years of age – unlike the Ets Chaim institution which catered to "advanced students training to be Talmud scholars or members of the rabbinate" (Henderson p. 149). Henderson provides no details of the curriculum at the junior institution,[9] but one account that I have seen relating to the previous century indicates that at that time (1688) at least the elder boys in the fifth and final classes had some Talmudic training preparatory for entry into Ets Chaim (*The Jewish Encyclopedia*, XII, p. 39). Nonetheless, even if we take for granted that this remained the case in Ricardo's day – and presuming always that he attended the institution – we would still be unjustified to base a strong link of the kind that Henderson forges *without specific knowledge of the training Ricardo actually received*. I note in passing a revealing remark at the close of the first chapter: "Since David Ricardo was born and reared in the highly structured and traditional Sephardic culture, it is with a high degree of probability that we can speculate about his life in its early period, even when there is little evidence" (p. 49). A rather risky viewpoint, I would suggest. (I do not doubt for one moment that someone with Ricardo's sort of mind would have taken to the Talmud like a fish to water; but this is to reverse causality.)

My second point relates to Henderson's position that "Ricardo's great theoretical contribution" was his analysis of the labor-capital process of industrial production, and not the labor-land process of agricultural production" (p. 19), and that this analysis reflected a process of transition from a society dominated by the system of labor-land production to one characterized by labor-capital activity (pp. 43–44) – a highly novel contribution, for "when David Ricardo, after Waterloo, fashioned the economic theory of the new labor-capital economy, he seriously threatened the idyllic image of the labor-land system that had dominated eighteenth-century Britain. Ricardo's economics was not just a new system, but one that tore at the roots of all that was cherished and admired" (p. 114). Much stress indeed is placed on Ricardo as creature of the city – in contrast to Malthus with his agricultural bias (pp. 115–116) – indicated inter alia by evidence of an attraction to hustle and bustle:

> In his youth, David Ricardo had no firsthand knowledge of the life style of a village, since he never lived in one. In cities, where his critics saw greed and ambition run amok, Ricardo found the hustle and bustle of his youth and that was "gratifying." Moreover, his attitude as to the advantages of city life did not change, even after he moved to Gloucestershire. On his estate, Gatcomb Park, he had the opportunity to compare country and city life, but it is clear that he always preferred the latter (p. 119).

After reference to Ricardo's roots in (financial) London and Amsterdam (p. 121), Henderson spends some fourteen pages on "the city and metropolis of London" and "the London Mob," as background to "the Parental Family," where we read that "[s]teeped in financial institutions, the Sephardic Jews of England were associated with the ever-increasing influence of finance capital" (p. 141).

Now I can appreciate that Ricardo's early years as "apprentice" to his father and the general commercial environment of his family is *one* of the considerations that may be relevant, in some manner, in approaching aspects of the early contributions to the bullionist debates or the currency pamphlets of 1816 and 1824 (p. 370). The problem is that we also find attributed to Ricardo, at least in 1814 and 1815, *corn-profit reasoning*:

> In the *Essay*, Ricardo had attempted to demonstrate that, with diminishing returns in agriculture, the rate of "surplus produce" would decline as accumulation took place. By showing that a fall in the facility of producing corn would raise the proportion of corn required as an input to produce a given corn output, he argued that a rise in wages, or the corn input, was necessarily accompanied by a fall in the rate of profits, when profits were viewed as a "deduction" from total output (p. 432).

Again: "In the *Essay*, Ricardo had used agricultural profit, expressed in terms of a corn input-output ratio, to measure changes in profits throughout the system when wages rose," and here Sraffa is cited (1951, p. xxxii) on the advantages of this method (p. 435). This corn-profit attribution is represented as typically "physiocratic:"

> Ricardo never mentioned the physiocratic notion of a net product, although Malthus referred a number of times to such a concept during the course of their debate in 1814 (*Works*, Vol. IV, p. 26). But because he relied upon Malthus's concept of rent, Ricardo implicitly showed an affinity for the physiocratic notion that the surplus output on land was the first regulator of profits. At this particular time in the formulation of his theory of profits Ricardo had not worked out the details of his theory of value, and his presentation was similar to that of the surplus theory of Quesnay in the *Tableau* (p. 330).[10]

And Henderson proceeds to cite Marx on the Physiocrats, interpolating the alleged application to Ricardo:

> Their [his] method of exposition is, of course, necessarily governed by their [his] general view of the nature of value, which to them [him] is not a definite social form of existence of human activity (labor), but consists of material things—land, nature, and the various modifications of these material things.
>
> The difference between the *value* of labor power and the *value created by its use* ... appears most tangibly, most incontrovertibly, of all branches of production, in agriculture, primary production. The sum total of the means of subsistence which the worker consumes from one year to another, or the mass of material substance which he consumes, is smaller than the sum total of means of subsistence which he produces ... In agriculture ... [this] shows itself directly in the surplus of use values produced over use values consumed by the worker, and can therefore be grasped without an analysis of value in general, or a clear understanding of the nature of value. This is true even when value is reduced to use value, and this latter to material substance in general. Agricultural labor is therefore for the Physiocrats [and Ricardo in the *Essay*] the only productive labor, because it is the only labour [Ricardo would say regulator] which creates surplus value, and *land rent is the only form of surplus value* which they [he] recognizes (p. 331).

Now if we opt for the high significance of environmental influence, such a model – which I myself do not find in Ricardo – would surely be far more likely to spring from the mind of someone steeped in the "*old*" world – Malthus perhaps?[11] It is a pity that Henderson did not notice and reflect on the serious implications of this case study for his general argument.[12] For it seems to me that only by tests such as the foregoing can one hope to validate the fruitfulness of biography. Professor Samuels cites a remark by Schumpeter on Keynes's *Essays on Biography* regarding "[t]he great difficulty of a biography like this [that] consists in making one connected whole of its two elements, disposition of a life and exposition of scientific achievements, which are so refractory to being welded together" (p. 12). For the reasons alluded to above, I am not convinced that Henderson overcame the obstacles to so refractory an endeavor.

VII

Professor Henderson drew upon the psychoanalytical literature, particularly the works of Erik H. Erikson, to assist in his interpretations. Consider the proposition that "David Ricardo's life shows a remarkable conformity to Erikson's life-cycle. There was an identity crisis, at the time when he broke with his parents, left the Sephardic enclave, married outside the faith, and entered upon his own business career" (p. 46; also on Erikson see pp. 155, 161, 177, 269). What identity crisis? Nowhere in his book does Henderson indicate evidence of any features of "crisis." I can think of no one less prone to personal crises than Ricardo – precisely the Ricardo that in fact emerges from Henderson's pages. Indeed, to the list of positive achievements outlined in Section IV above I would certainly add the portrait painted of the man, above all his generosity – though Henderson actually takes him to task for *excessive* kindness: "there was also the personality characteristic" (referred to on p. 190 as a "compulsion") "which fostered the practice of giving money away" (pp. 370–371); his fairness, including fairness in debate, and (what is related) his quest after truth rather than personal intellectual priority; and his integrity. Henderson believed that Ricardo's kindness and generosity were "virtues inherited from the cultural tradition of the Sephardic enclave," that it was "this heritage which motivated him to be generous and free with his money" (p. 190); and that he derived from his heritage a concern with the "public interest" and from his father, his high sense of integrity (p. 370). Perhaps so. One would certainly like to think so.

VIII

In summary, it has emerged that Professor Henderson and I are in love with the same *man*, but not with the same *economist*, a formulation that Henderson, of course, would not have accepted. For Henderson's portrayal of the man I am grateful. The book is a worthy tribute to its author, especially if the errors of fact can be corrected for any second edition and an adequate Index prepared. And it may well be that several of what I consider more serious "defects" reflect nothing more than a reaction to a different approach towards intellectual history than my own.

I close by expressing a particular obligation to Professor Henderson for bringing to my attention the fact that the Jewish birthdates of Ricardo and myself are close enough to have assured that we read from the same Bar Mitzva Sedra (see Henderson on Ricardo, p. 154), namely Leviticus 9–11. This is very pleasing to me. (Why this should be so must be left for further introspection on my part.) Were Henderson still with us I would repay my debt by letting him know that, several years ago, I came across a document indicating that long after his quitting the Synagogue, Ricardo made a financial contribution to it. I suspect that we have here nothing more than a further illustration of Ricardo's generosity, but Henderson would surely have been delighted with this addition to our knowledge of the man.

NOTES

1. Henderson cites Hyamson, 1951 on the constitution of the Synagogue which requires there to be no distinctions "whether in respect to the graves, or the honors conferred in Synagogue" (p. 52). This is said of the seventeenth-century but the theme is applied more generally.

2. On various contrasts drawn between the Ashkenazi and Sephardic communities, see also pp. 24, 55, 141, 166.

3. In a recent memoir I document an instance in my own life of Ashkenazi intolerance towards Sephardim (Hollander, 1998a).

4. Of course, Henderson is not the first to see things like this (see in particular Marshall, 1920, p. 761n).

5. In the course of his discussion of Malthus, Henderson emphasizes his alleged authorship of the January 1808 article "Spence on Commerce" for the *Edinburgh Review* (see Henderson, pp. 225, 233, 235–237, 455). In my opinion, this paper was by Brougham not Malthus (see Hollander, 1998b). Malthus's authorship was also questioned by Patricia James (as Henderson notes in passing on p. 236).

6. He also expresses debts for biographical detail to Heertje, 1975; and, to a lesser degree, to Hasson, 1968 and Weatherall, 1976.

7. There is a little imprecision here, for the relevant section heading reads "The Partial Eclipse of Ricardian Theory" (p. 621), but the text refers simply to "the eclipse . . ." (p. 625).

8. In his Introduction, Samuels refers to Whig History as though it were specifically a neo-classical procedure (p. 3), but goes on to allow in effect that neo-Ricardians and Marxists also engage in Whig History and that Henderson does just this.

Henderson's tone throughout is moderate, but he was capable on occasion of using sharp epithets in dealing with those whose interpretations he disliked, e.g. "arrogance" for George Stigler and "muddled" and "ancient" for Lord Robbins (pp. 651, 654). I am, however, surprised to find some rather harsh remarks regarding Gramsci – "an exceedingly strident critic of Italian fascism" (p. 638n) – and even Marx who "let his family suffer" and "starve" (p. 45n).

9. The reader should be aware that the term "Talmud Tora" does *not* refer to study of the Talmud – the word "Talmud" in this context conveys simply the notion of "learning," and is the term typically used for *elementary* education.

10. Sraffa too pointed to a Ricardo-Physiocratic link (1960, p. 93). Also Sraffian is the insistence – I would say quite valid in this case – that the Ricardo of the *Principles* can only be understood if it is recognized that the purpose of the labor-embodied theory was to serve as a device to deal with the problem of distribution rather than as a mere theory of price; similarly, such recognition permits the proper evaluation of the

adjustments made in the 1821 edition (Chapter VIII *passim*; pp. 496, 646). This feature of Henderson's work is particularly emphasized by the editor (Introduction, p. 7).

11. I myself *do* attribute a corn-profit model to Malthus and relate it to physiocratic bias (Hollander, 2000).

12. One might answer on Henderson's behalf that Ricardo after all abandoned the corn-profit model once his thinking had "undergone modification and development" (p. 331), and that the process of modification points away from agriculture. I would be prepared to counter any such argument on the grounds that it renders the "environmental" feature far too loose to be helpful.

I note here as an aside a problematic assertion by Henderson relating to the significance of the contemporary transition of the economy: "The scientific aspects of Marx's socialism were grounded in English classical political economy, a political economy grounded in the material conditions of English society in the eighteenth century" (p. 180). This statement regarding the *agricultural roots* of English classical political economy, is quite at odds with all that Henderson has written on Ricardo.

REFERENCES

Gaster, M. (1901). *History of the Ancient Synagogue of the Spanish and Portuguese Jews . . . Bevis Marks*. London.

Heertje, A. (1975). On David Ricardo (1772–1823). *The Jewish Historical Society of England, Transactions*, Vol. XXIV & Miscellanies, Part IX, pp. 73–82.

Hasson, J. A. (1968). David Ricardo – Sephardic Genius. *American Society of Sephardic Studies* (a Yeshiva University journal), I, pp. 93–112.

Hollander, S. (1998a). It's an Ill Wind . . . A Memoir. In: *The Literature of Political Economy: Collected Essays II* (pp. 3–27). London: Routledge.

Hollander, S. (1998b). On the Authorship of 'Spence on Commerce.' In: *The Edinburgh Review*, January 1808," in *The Literature of Political Economy: Collected Essays II* (pp. 349–362). London: Routledge.

Hollander, S. (2000). Malthus and the Corn-Profit Model. In: H. Kurz (Ed.), *Critical Essays on Piero Sraffa's Legacy in Economics* (pp. 198–222, 246–256). Cambridge: Cambridge University Press.

Hyamson, A. M. (1951). *The Sephardim of England*. London.

The Jewish Encyclopedia (1916). 12 volumes, New York and London: Funk and Wagnalls Company.

Marshall, A. (1920). *Principles of Economics* (8th ed.). London: Macmillan.

Sraffa, P. (1951). Introduction to *Works and Correspondence of David Ricardo*, vol. I, Cambridge: Cambridge University Press.

Sraffa, P. (1960). *The Production of Commodities by Means of Commodities*. Cambridge: Cambridge University Press.

Torrens, R. (1844). *The Budget: On Commercial and Colonial Policy*. London: Smith, Elder.

Weatherall, D. (1976). *David Ricardo: A Biography*. The Hague.

Hollander's THE ECONOMICS OF THOMAS ROBERT MALTHUS

REVIEW BY JEFFREY T. YOUNG

Toronto: University of Toronto Press, 1997. Pp. xviii, 1053.

That Robert Malthus would easily win the prize for most commonly misunderstood economist may be seen immediately from the following comment found in J. S. Mill's *Principles*:

> The publication of Mr. Malthus' *Essay* is the era from which *better views* of this subject (the relation of people to food) must be dated; and notwithstanding the acknowledged errors of his first edition, few writers have done more than himself, in the subsequent editions, to promote these *juster and more hopeful* anticipations. (1987 (1848), p. 747, emphasis added)

Despite James Bonar's very creditable attempt over a century ago to rescue Malthus, his name seems to be irrevocably associated with gloom and doom.[1] (Bonar, 1924 (1885))

This may now be changing, at least among the experts. If not exactly a flood, there have been a series of important studies on Malthus within the last two decades, and in 1998 the top specialist journal in the history of economic thought devoted over 70 pages to the assessment of these developments in Malthus interpretation. (Waterman, 1998; Hollander, 1998a; Winch, 1998; Pullen, 1998) New ground is being broken in our understanding of the intellectual history of the classical era in Britain in which Malthus emerges as

Research in the History of Economic Thought and Methodology, Volume 19A, pages 297–310.
2001 by Elsevier Science B.V.
ISBN: 0-7623-0703-X

a key figure in the attempt, by mostly Christian writers, to maintain political economy as a moral science. In addition, Malthus's difficult and often confusing analytical material has been the subject of rational reconstructions that enhance Bonar's judgment that as a theorist Malthus was Ricardo's equal. (1924, v.)[2]

Notwithstanding these achievements, Hollander's *The Economics of Thomas Robert Malthus* (hereinafter referred to as *ETRM*) is by far the most ambitious and definitive treatment of Malthus to date. It is destined to become the standard reference on Malthus's economics. Stylistically the book is isomorphic with his previously published works. It is long (1007 pages), closely reasoned, exhaustive in its coverage, and quotes extensively from Malthus's original text. This volume is more user friendly, however, than its immediate predecessor on J. S. Mill. There is a fifty-plus page concluding chapter that brings together all the important conclusions of the book that may be usefully read independently of the book itself.[3]

Despite its encyclopedic nature there is a strong storyline here. Hollander's method of meticulously sifting through everything an author wrote in the manner of a Biblical commentary has produced new and controversial insights. What ultimately emerges from these pages is a new, fascinating, and controversial interpretation that portrays Malthus as caught between two opposing approaches to economics. These are the scarcity approach of Ricardo and the surplus approach of the physiocrats. The story of Malthus's development as a theorist is a dramatic tale in which the former (scarcity and allocation economics) gradually achieves ascendancy over the latter (surplus economics), culminating in Malthus's recantation of his position on agricultural protectionism. In the end, the conversion is incomplete as Malthus continues to oppose Say's Law orthodoxy.

We may gain an appreciation of this story by examining three interrelated themes. First, and perhaps the most controversial and newsworthy, is Hollander's claim (previously announced in the prestigious pages of the *American Economic Review*) that sometime around 1824 Malthus renounced his stand on agricultural protection and threw his support to the Ricardian vision of Britain's economic future fueled by industrial growth coupled with free trade in both corn and manufactured goods. (Hollander, 1992) The upshot is that eventually Malthus joined the classical mainstream with the only substantive point of disagreement being his rejection of Say's Law which survived his renunciation of agricultural protection even though effective demand considerations was one of the arguments he had used to support a high rent, self-sufficient agricultural sector.

Second, Hollander argues that there is a strong physiocratic influence on Malthus, which waned with his reversal on the Corn Laws. The presence of an agricultural bias standing alongside such classical commonplaces as the law of diminishing returns Hollander treats as evidence for the two conflicting paradigms view of Malthus. This tension becomes the key to the problem of Malthus's consistency. Hollander does not raise directly the issue of the single versus dual paradigms in the history of economic thought that played such a prominent role in his *Economics of David Ricardo* (hereinafter *EDR*) and its critical reception.

Third, closely associated with the physiocratic surplus approach, is Hollander's claim that Malthus adopted the Sraffian corn-model theory of profits along with a Sraffian type of price theory, which historians have usually attributed to Ricardo. Since Malthus's explicit discussion of the corn-model theory in the second edition of his *Principles* has been cited as evidence in support of Sraffa's interpretation of Ricardo, Hollander's claim about Malthus is likely to be equally controversial as his interpretation of Ricardo (Prendergast, 1986). In this way *ETRM* reinforces the argument of *EDR*. In what follows I will attempt to present the substance of Hollander's case on each of these points.

Corn Laws

With regard to the Corn Laws, Hollander's evidence comes from four separate sources dated from 1824 until Malthus's death. They are the nature of certain changes made in the second edition of the *Principles*, an 1824 *Quarterly Review* article on Ricardian political economy, a note added to the chapter on import restrictions for corn in the 1826 edition of the *Essay on Population*, and three letters written between 1829 and 1833.

Space does not permit a thorough review of all the evidence, but we can gain an appreciation of the case by looking at the note added to the *Essay*. The note is long, and it begins with a refutation of one common argument favoring free trade in corn. Among other things the opponents of the Corn Laws claimed that free importation would protect Britain from the distress of a scarcity of corn in the home market. Malthus points out that this may not be the case since the facts point to similar weather conditions in all of the European corn growing countries causing good and bad harvests to occur simultaneously throughout Europe. Free trade, then, could not be relied upon to stabilize prices and available quantities. At this point, instead of his usual endorsement of the Corn Laws as a necessary exception to the policy of free trade, Malthus states:

I am very far, however, from meaning to say that the circumstances of different countries having often an abundance or deficiency of corn at the same time, though it must prevent the possibility of steady prices, is a decisive reason against the abolition or alteration of the corn-laws. The most powerful of all the arguments against restrictions is their unsocial tendency, and the acknowledged injury which they must do to the interests of the commercial world in general. The weight of this argument is increased rather than diminished by the numbers which may suffer from scarcity at the same time. And at a period when our ministers are most laudably setting an example of a more liberal system of commercial policy, it would be greatly desirable that foreign nations should not have so marked an exception as our present corn-laws to cast in our teeth. A duty on importation not too high, and a bounty nearly such as was recommended by Mr. Ricardo, would probably be best suited to our present situation, and best secure steady prices. A duty on foreign corn would resemble the duties laid by other countries on our manufactures as objects of taxation, and would not in the same manner impeach the principles of free trade.

But whatever system we may adopt, it is essential to a sound determination, and highly useful in preventing disappointments, that all the arguments both for and against corn-laws should be thoroughly and impartially considered; and it is because on a calm, and, as far as I can judge, an impartial review of the arguments of this chapter, they still appear to me of weight sufficient to deserve such consideration, and not as a kind of protest against the abolition or change of the corn-laws, that I republish them in another edition (1992 (1826), p. 180).

The allusion to the "present situation" suggests that at least part of what is going on is that Malthus has changed his perception of the balance of costs and benefits associated with free trade, because circumstances had changed. As Hollander points out, by 1826 the liberalization of trade was becoming a reality rendering obsolete Malthus's earlier position which was in part based on the belief that free trade was an unrealizable ideal (*ETRM*, p. 857; also 1992, p. 653). Malthus never disputed that free trade was right in principle. He always viewed the Corn Laws as a legitimate exception to a policy of free trade in all other sectors of the economy.

In addition, Hollander argues that the endorsement of Ricardo's plan for export subsidies is actually an endorsement of free trade, not the reverse, as a superficial reading would imply. Ricardo's plan called for a duty on corn imports ". . .to correct for differential taxation imposed on British farmers relative to British manufacturers, and an appropriate drawback on (corn) exports" (*ETRM*, p. 851). The purpose was to prevent the distortion of resource allocation that resulted when imported corn was not subject to the same taxation as homegrown corn. Despite the ambiguous way Malthus refers to Ricardo's scheme (are both the "duty" and the "bounty" meant to refer to Ricardo or just the "bounty"?), an endorsement of the whole policy package seems to be the most logical interpretation, since Ricardo's proposed drawback does not stand alone as a meaningful policy proposal (*ETRM*, p. 853).

Also worth noting is that Malthus himself tells us why he did not change the body of the text to reflect his latest thinking on agricultural protection, thus relegating the "announcement" of his new position to a footnote. The importance of the issue made it incumbent upon political economists to give a full airing to all the arguments. In addition, he continued to believe that protectionist arguments were sufficiently meritorious to be listened to and duly considered in the public arena. He specifically did not retain the protectionist passages as a "protest" against the abolition or even change of the Corn Laws.

Hollander's case is by no means airtight. There is the thorny problem that two of Malthus's close friends, John Cazenove and William Empson, appeared to know nothing of his conversion, and that the second, posthumous edition of the *Principles* retains a formal statement of the desirability of agricultural protection. (Pullen, 1995, pp. 522–523) Given the unfinished nature of the second edition when Malthus died, Hollander discounts the significance of this. The latter parts cannot be taken at face value to represent Malthus's mature views. In addition, Malthus was not known to be forthright and open when he did change his mind.[4]

Surplus vs. Scarcity

Important analytical issues are also at stake in the question of Malthus's stand on the Corn Laws. An important attribute of Hollander's reading of Malthus is the presence of a physiocratic bias, more far-reaching and ingrained than Smith's, that played a key role in his original stance in favor of the Corn Laws, and which Hollander claims waned with his recantation in the mid-1820s.

That the first two editions of the *Essay on the Principle of Population* contained strong physiocratic influences is well known. Hollander's claim to originality here is that he sees this influence as ". . . more positive, specific, and lasting than usually suggested in the literature" (*ETRM*, p. 406). Winch's assessment is perhaps typical in this regard when he labels the physiocratic features of the *First Essay* a "flirtation" and concludes that they were ". . . largely excised three years later" (1987, p. 61). Particulars include a physiocratic definition of wealth wherein manufacturing income is treated as a transfer payment, agriculture as the sole source of surplus, the view of rent as derived from a Providential gift, the greater productivity of a unit of capital in agriculture compared to either trade or manufacturing, and landlord expenditures as the determinant of aggregate demand. By 1806 Malthus was treating manufacturing and trade as creators of income, but not producers of surplus. Thus, the agricultural surplus was still seen as the source of all forms of disposable income, i.e. income available for either taxation or accumulation without causing a withdrawal of productive services.

The result was that Malthus continued to look upon agriculture as a special activity. It tended to yield a social rate of return in excess of the private rate (the opposite applied to manufacturing). It also had the special property that, unlike the market for other goods where supply and demand operated as independent functions, supply created its own demand (*ETRM*, p. 381). An increase of mouths to feed was sure to follow an increase in food output guaranteeing that the physical corn surplus would always result in a value surplus to be distributed as rent and profits to farmers and manufacturers. Only in the case of corn did Malthus attribute this value surplus to the physical surplus of output above capital consumed (*ETRM*, p. 379).

Perhaps Malthus's most significant departure from physiocracy was the principle of diminishing returns that formed the basis of the classical theory of rent, of which Malthus, of course, was a principal founder. At the same time this introduces a scarcity perspective which is at odds with the "surplus" approach. This is reflected, for example, in Malthus's and Ricardo's disagreement over the nature of rent. In Ricardo's view rent arose out of the scarcity of fertile land, and was larger the more severe the scarcity relative to the demand for food (*ETRM*, p. 380). Rent did not reflect a creation of wealth, only that the natural agents, which cooperated with man in agriculture, were scarce, while in manufacturing they were not. A low rent, high profit, and high wages economy was to be preferred over the high rent economy. Malthus saw rent as a share of the agricultural surplus, a gift of Providence. To him, high rent and high wages went together since a higher money price of corn would increase both real rent and real wages. This economy was to be preferred over a high profit economy. The high rent, high wage economy could support a larger population, and the consumption propensity of landlords would insure macroeconomic stability. Malthus recognized land scarcity as a condition for rent to exist, but viewed it as only determining the distribution of the surplus as rent (*ETRM*, p. 389).

Malthus did seem to be aware that he was trying to reconcile competing doctrines. In addition to the tactic of treating scarcity and surplus as two separate and necessary causes of rent, he also suggested that the surplus approach should be confined to the special case of necessaries, all other goods governed by relative scarcity via the mechanism of independent supply and demand functions (Hollander, 1998, p. 302). However, Malthus does not seem to have been aware that he was dealing with mutually exclusive paradigms of pricing and distribution. The superior productivity of agriculture simply cannot be sustained in the face of diminishing returns in that sector. At the margin, labor must at some point become more productive in manufacturing than in agriculture. Moreover, he could not maintain two principles of price determination, both applicable to agriculture. Corn prices could not simultaneously be determined by the intersection of an upward sloping long-run supply curve with a downward sloping demand curve leaving rent as a payment arising out of the scarcity of land (which Malthus upheld) and also be sure that price would always return a value surplus to the farmer since supply created its own demand. In the first instance the supply and demand functions are independent, as they must be, while in the second they are interdependent. Hollander concludes:

> He had fallen into a logical trap – insisting upon the physiocratic view of rent as surplus while allowing the differential principle of rent. Malthus's theoretical position was, as Ricardo insisted, untenable (*ETRM*, p. 398).

A further implication of the doctrine of the unique productivity of agriculture is that the unregulated market allocation of resources between agriculture and manufacturing is likely to be non-optimal from the social perspective. Indeed, one wonders if the physiocrats could really have been as committed to free trade in principle as they are generally portrayed (today as well as in Malthus's time).[5] In Malthus's case there is a clear association between physiocratic principles and agricultural protection. The agricultural origin of the surplus renders agriculture more productive per unit of capital invested than manufacturing, and since the surplus does not accrue exclusively to the farmer, the social rate of return in agriculture exceeds the private rate (Hollander, 1998b, p. 290, 299). In so far as markets respond to private calculations of gain, the result is that capital will be under-invested in agriculture and over-invested in manufacturing even if the restrictions, which favor the latter over the former, are removed. Subsidies and import restrictions are called for on utilitarian grounds.

In addition, there are aggregate demand considerations. Malthus claimed, against Ricardo and his school, that capital accumulation was not alone sufficient to sustain growth, and that diminishing returns is not the only inhibiting factor. In particular, he believed that a strong demand for leisure would stunt the growth of effectual demand unless new wants were constantly being created (*ETRM*, p. 581). This was one of the functions of trade in his theory. Expansion of aggregate demand was for him a gain from trade separate from and additional to the efficiency gains of Ricardo's model. In addition, he feared that investment derived from a reduction in consumption would be unsustainable since there would be an insufficient market for the expanded supply of goods. In short, he feared that if the propensity of landlords to save was too high the capitalists' excess of their investment over their saving was unlikely to be sufficient to bring total investment in line with total saving, thus generating an insufficiency of aggregate demand. This was a long-run problem with the secular growth path, not the Keynesian short-run problem, which Malthus also analyzed.

With landlord marginal propensity to consume being generally higher than that of the capitalists, a high rent share would tend to alleviate this problem. Hence, agricultural protection was in part advocated to increase aggregate demand and encourage sustainable accumulation (investment financed out of rising profits). This was his version of the model of balanced growth. Over time the agricultural and manufacturing sectors of the British economy would grow together. There would be no secular shift toward manufacturing as the principle of comparative advantage entailed if allocation were to be efficient. Malthus's position on aggregate demand is more complex than this, as we have made no allowance for short-run fluctuations and the analysis of the postwar depression in Britain. However, the connection between agricultural protection, physiocracy, and demand seems to lie in his analysis of the role of demand as a causal factor in generating the secular growth path.[6]

Over time Malthus's physiocratic bias weakened. Although he never completely relinquished it, he did eventually endorse the Ricardian vision of industry-led growth in a free trade regime.[7] The process began with the introduction of prudential control into his population model, which tends to undermine the "surplus" approach, as the wage now becomes an endogenous variable set by the operation of supply and demand (Hollander, 1998b, p. 310). This blurs the distinction between surplus and necessities. By the 1817 edition of the *Essay* he was beginning to place an emphasis on industry-led growth, and he eventually abandoned the notion that landlord consumption was necessary to maintain aggregate demand (1998b, pp. 310–311). This latter derived from his reading of the experience of British growth in the 1820s in which growth in manufacturing exports was clearly the engine of aggregate demand growth. Although he abandoned protection and embraced much of the Ricardian vision, he never relinquished his claim against the Ricardians that aggregate supply growth did not guarantee aggregate demand growth, or that a breach between ex ante saving and investment could temporarily interrupt the secular growth process.

Malthus has suffered from bad press in the history of economic thought, possibly because of the tendency for the victors to write the history. Keynes famously tried to redress the balance and refurbish Malthus's reputation. However, there has been a tendency to view him as confused, inconsistent, or even worse. Hollander's work sheds some important light on why Malthus has this reputation. His thought evolved as he grappled with the demands of two conflicting approaches. To make matters worse he was trying to develop a model that was inherently more complex than that of his contemporaries as he attempted to graft an aggregate demand function onto the corpus of canonical classical theory. It is not, therefore, surprising that Malthus should gain such a reputation. *ETRM* goes a long way to offering a plausible interpretation of Malthus that does not rely on assuming that Malthus was a theoretical lightweight compared to Ricardo.

Sraffian Themes

There remains the issue of Malthus's Sraffian-style value and distribution theory that is also associated with his physiocratic bias, the surplus approach. Hollander is not the first to suggest that Malthus followed a Sraffian approach, but there is a difficult matter of interpretation here (Costabile, 1983). On the one hand, there is a clear Sraffian model in Malthus with value determined by quantities of dated labor, the profit rate determined as the ratio of surplus to capital advanced, and the labor commanded value measure. This is not in dispute between Hollander and the Sraffians. On the other hand, passages in which Malthus is appears to be criticizing the idea of profit as a physical ratio have been used as evidence in support of Sraffa's interpretation of Ricardo (Prendergast, 1986). This is where the problem lies. Hollander asserts that it is Malthus, not Ricardo, who spelled out the physical corn profit model with manufacturing profits coming into line via the adjustment of the corn/cloth terms of trade and distribution determined prior to prices (*ETRM*, p. 435, *EDR*, p. 722). In the multi-goods case the rate of profit is determined in the wage goods sector. The Sraffians dispute this arguing that Malthus's theory derives the profit rate as a value ratio in the manner of *Production of Commodities*. The interpretive issues are complex, but we may at least gain an appreciation of Hollander's position.

Malthus always defined the profit rate as a value ratio. However, Hollander points out that

> ... though the general expression for the profit rate involves value terms, the *agricultural* profit rate is expressed in physical (corn) terms, and the rate thus determined carries over to manufacturing. Second, Malthus spelled out the assumptions required to permit proceeding in physical terms in agriculture; and he cautioned that, in application, the procedure might break down, even should both input and output be composed of the same physical substance. The evidence points: (1) to an *analytical* corn-profit model consciously spelled out ... and (2) to limitations in application reflecting largely seasonal corn-price fluctuations (*ETRM*, p. 436; emphasis in original).

We have here the substance of Hollander's interpretation reduced to two propositions. First, there is a consciously spelled out corn-profit model to be found in both editions of the *Principles*. The evidence is the use of physical measures of input and output in agriculture. Second, there is an explicit rendering of the assumptions necessary for the physical ratio to determine the value ratio. In the scenarios where Malthus uses these assumptions, the lines of causation are from agriculture to manufacturing via the adjustment of the relative price of manufactured goods.

One such example is his exposition of the canonical classical model of growth under conditions of increasing land scarcity in the second edition of the *Principles*. After showing that diminishing returns lowers profits and corn wages, Malthus queries:

> In the meantime, it will be asked, what becomes of the profits of capital employed in manufactures and commerce, a species of industry not like that employed upon the land, where the productive powers of labour necessarily diminish; but where these powers not only do not necessarily diminish, but very often greatly increase? ((1836) 1986, pp. 274–275)

He goes on to answer that

> In the cultivation of land, the cause of the *necessary* diminution of profits is the diminution in the quantity of produce obtained by the same quantity of labour. In manufactures and commerce, it is the fall in the exchangeable value of the same amount of produce (p. 275; emphasis in original).

He then explains, using his definition of value as labor commanded, how capital will flow into manufacturing until the value of manufactured goods falls sufficiently to reduce the manufacturing rate of profits to equality with agriculture. From the context it is clear that the agricultural profit rate is ultimately a physical phenomenon involving the relation of corn input (corn wages) to corn output. Consequently the agricultural rate is the governing rate for the system. This is precisely the model that Sraffa attributed to Ricardo (*ETRM*, p. 447).

Second, Malthus recognized limits to the applicability of the model. As a model of the secular path of wages and profits, it incorporates only the impact of diminishing returns in a world of constant technology. Malthus, of course, was never satisfied that diminishing returns told the whole story. It constituted what he called the "limiting" principle, while the impact of aggregate demand and supply were the "regulating" principle (*ETRM*, p. 448; Malthus (1836) 1986, p. 271). If aggregate demand were insufficient, as in a depression or secular stagnation, the rate of profits would fall below that dictated by the marginal product of capital advances in agriculture. Seasonal fluctuations in harvests could also cause the actual rate of profit to deviate from that determined at the margin in agriculture. Short-run supply variation, for example, can cause price fluctuations that affect input and output values differently, even though both are corn. It is considerations such as these which led Malthus to insist that profits are always a value ratio, but this does not rule out a corn-model since, as Malthus claims, the value ratio is essentially determined by the interaction of the "limiting" and "regulating" principles. The "limiting" principle is a Sraffian model with profits determined in the wage goods sector independently of, and prior to, prices.

With his recantation of agricultural protectionism, its associated distancing from physiocratic principles, and his ascent to the inverse wage/profit relation, Malthus moved closer to Ricardianism in the 1820s. However, there remained three essential differences between them; two are substantive and one is erroneous. Malthus erroneously believed that matters of substance depended upon his choice of numeraire, labor commanded, which necessarily rendered the wage basket per unit of labor constant in value. However, as Hollander shows there is no substantive difference here. Diminishing returns generates the same secular trends regardless of whether one uses Malthus's or Ricardo's measuring stick (*ETRM*, pp. 500–501).

The substantive differences are Malthus's use of the corn-model to explain the "limiting" principle of profits and his theory of aggregate demand. In the first instance, the key is that while Malthus assumed the money wage was constant (Malthusian "money", unlike Ricardian "money", was produced by a constant amount of unaided labor in one hour of work), money wage variation was the source of profit rate variation in Ricardo's model (*ETRM*, p. 501). The result is that Ricardo did not use a corn-model (although he could have used it as a special case) while Malthus did use it. In the second instance, Malthus once again failed to realize that his doctrine did not depend on his choice of numeraire. One does not need a labor commanded index to arrive at the conclusion that aggregate supply can exceed aggregate demand. The upshot is that there are indeed differences between the mature Malthus and Ricardo, but they are not as great as Malthus himself believed.

CONCLUSION

This volume has and will raise controversy. As a companion to *EDR* it reinforces Hollander's iconoclastic reading of Ricardo. However, like the rest of Hollander's *oeuvre* it is so thoroughly researched and carefully and closely argued that it must be taken extremely seriously. It will undoubtedly produce converts, as has Hollander's view of Ricardo, while immensely irritating various entrenched orthodoxies. Foremost among these is the Sraffian view of classical economics. Presenting Malthus as more Sraffian than Ricardo ever was adds a new and ironic twist to what is now an old debate, suggesting that

Sraffa along with Keynes should have lamented the triumph of Ricardianism in the nineteenth century.

With reference to that debate the present volume invites the following set of hypotheses. First, the corn-model that Sraffa saw in Ricardo's "Essay on Profits" of 1815 is actually in Malthus, although it does not appear before the first edition of the *Principles* in 1820. Second, the error of attributing it to Ricardo was originally Malthus's error, and Sraffa was simply following Malthus. Indeed, it is a curious fact that much of the weight of evidence cited for the corn-model interpretation of Ricardo rests on what Malthus said. Third, although this is quite speculative because we do not know if Malthus had the corn-model in 1815, Malthus could easily have made this mistake if he was himself thinking in terms of the corn-model as a theory of the agricultural profit rate. If so, he would have naturally taken Ricardo's assertion in the "Essay" of the leading role of agriculture as an expression of the model.

Hollander draws other significant conclusions from his view of Malthus trying to square the circle of operating simultaneously within two mutually exclusive paradigms. On methodology he argues there is no significant difference between Ricardo and Malthus, and on Malthus's theory of effective demand he finds significant differences with Keynes, some of which may be traced to the physiocratic elements in Malthus. Ultimately what emerges is a very complex Malthus, not necessarily a confused, inconsistent, or theoretically inept one. It is also a Malthus whose thought is in continuous flux. Coupled with his "less than candid" way of changing his mind, it is not surprising that he presents a particularly difficult subject of interpretation. Hollander's monumental work may not be the last word, but it certainly deserves to set the standard for quite some time to come.

Jeffrey T. Young
St. Lawrence University

NOTES

1. In light of David Levy's recent, important discoveries, I will not implicate Malthus with the "dismal science" (2000). Keynes's famous attempt in the *General Theory* to recast Malthus as would-be hero also failed to reverse the image of Malthus as a lesser light compared to the Ricardian sun.

2. Particularly worthy of mention are Waterman's *Revolution, Economics, and Religion* and Winch's *Riches and Poverty* (1991 and 1996, respectively; also Winch's *Malthus*, 1987). These two excellent studies approach the subject from what Waterman calls the intellectual history perspective. (1998, p. 303) Both present Malthus as a Christian moralist, who was by no means pessimistic about long-run growth prospects, given prudential behavior. Waterman is also responsible for significant rational

reconstructions of Malthus's population theory. (see also Eltis, 1980; Costabile, 1983; and Costabile & Rowthorn, 1985) Hollander's work generally falls into the history of economic analysis perspective, and this volume is no exception. Although he objects to any suggestion that an airtight division exists between the two types of history (1998a, p. 341), this volume on Malthus may be thought of as complementary to the work of Winch. Certainly Hollander is not hostile to intellectual histories, and in his thoroughness he does take up the parts of Malthus's texts which form the substance of this approach. Contrary to Waterman and Winch, Hollander concludes that Malthus's theology has no bearing on his pure analysis. (*ETRM*, p. 918)

3. As a reviewer I am quite sensitive of my role in trying to tell the reader whether he or she should read the book under review. In this case, we have probably the best secondary work ever written on Malthus, but it is one which only a small group of aficionados will ever read from cover to cover. Others should approach this as they would a good encyclopedia. The chapters may be read as self-contained units with great profit. Combined with a reading of the conclusion one is unlikely to miss the thread of the story.

4. Waterman, for example, points this out in relation to the question of what happened to Malthus's theodicy after he had excised the last two chapters of the first *Essay* when he calls Malthus's method of recantation "less than candid." (1991, p. 146; see also pp. 138, 172). The citations that Hollander makes to Malthus's *Quarterly Review* article would not be viewed as a recantation unless they were set against the lengthy footnote discussed above.

5. In fact, Hollander holds the view that they were not committed to the doctrine of free trade in principle, regardless of the resulting resource allocation. Their advocacy of free trade in grain in France in the eighteenth century turned on their presumption that under such a régime France would be a net exporter of food, causing resources to be redirected from trade and manufacturing to agriculture (1987, p. 51.).

6. On the issue of short-run fluctuations, Malthus recognized the possibility of *ex ante* saving exceeding *ex ante* investment, thus bringing about an excess of aggregate supply over aggregate demand. However, in 1830 J. S. Mill showed that this condition was perfectly consistent with Say's Law in that this excess supply of commodities would be counter-balanced by an excess demand for money balances (Mill, 1830; *ETRM*, p. 629). On the supposed Keynesian flavor of Malthus's position Hollander notes that there were several differences. On public works, for example, Malthus thought they could help because they constituted unproductive expenditure. Keynes thought they brought about a "net injection of purchasing power" (*ETRM*, p. 629). Malthus might have stimulated Keynes, but Keynes produced a non-Malthusian theory of effective demand.

7. Although not going so far as Hollander, Winch has also noted in his writings that post-1820 Malthus did become more favorable to manufacturing. (1987, p. 65)

REFERENCES

Bonar, J. (1924 (1885)). *Malthus and His Work*. New York: Macmillan.

Costabile, L. (1983). Natural Prices, Market Prices and Effective Demand in Malthus. *Australian Economic Papers*, 22, June, 144–170.

Costabile, L., & Rowthorn, B. (1985). Malthus's Theory of Wages and Growth. *The Economic Journal, 95*, June, 418–437.

Eltis, W. (1980). Malthus's Theory of Effective Demand and Growth. *Oxford Economic Papers, 32*(1), March, 19–56.

Hollander, S. (1979). *The Economics of David Ricardo (EDR)*. Toronto: University of Toronto Press.

Hollander, S. (1987). *Classical Economics (CE)*. Oxford: Basil Blackwell.

Hollander, S. (1992). Malthus's Abandonment of Agricultural Protectionism: A Discovery in the History of Economic Thought. *American Economic Review, 82*(3), June.

Hollander, S. (1997). *The Economics of Thomas Robert Malthus (ETRM)*. Toronto: Toronto University Press.

Hollander, S. (1998a). An Invited Comment on 'Reappraisal of "Malthus the Economist", 1933–1997'. *History of Political Economy, 30*(2), Summer, 335–342.

Hollander, S. (1998b). *The Literature of Political Economy, Collected Essays II*, and London: Routledge.

Levy, D. (2000). 150 Years and Still Dismal! *Ideas on Liberty*, March, pp. 8–10.

Malthus, T. R. (1986 (1836)). *Principles of Political Economy Considered with a View to Their Application* (2nd ed.). New York: Augustus M. Kelley.

Malthus, T. R. (1992 (1826)). *An Essay on the Principle of Population*, edited by Donald Winch. Cambridge: Cambridge University Press.

Mill, J. S. (1974 (1830)). *Essays on some Unsettled Questions of Political Economy* (2nd ed.). New York: Augustus M. Kelley.

Mill, J. S. (1987 (1848)). *Principles of Political Economy with Some of Their Applications to Social Philosophy*, edited by Sir William Ashley. New York: Augustus M. Kelley.

Prendergast, R. (1986). Malthus's Discussion of the Corn Ratio Theory of Profits. *Cambridge Journal of Economics, 10*(2), June, 187–189.

Pullen, J. (1995). Malthus on Agricultural Protection: An Alternative View. *History of Political Economy, 27*(3), Fall, 517–529.

Pullen, J. (1998). The Last Sixty-Five Years of Malthus Scholarship. *History of Political Economy, 30*(2), Summer, 343–352.

Waterman, A. M. C. (1991). *Revolution, Economics and Religion: Christian Political Economy 1978–1833*. Cambridge: Cambridge University Press.

Waterman, A. M. C. (1998). Reappraisal of 'Malthus the Economist,' 1933–97. *History of Political Economy, 30*(2), Summer, 293–334.

Winch, D. (1987). *Malthus*. Oxford: Oxford University Press.

Winch, D. (1996). *Riches and Poverty: An Intellectual History of Political Economy in Britain, 1750–1834*. Cambridge: Cambridge University Press.

Winch, D. (1998). The Reappraisal of Malthus: A Comment. *History of Political Economy, 30*(2), Summer, 353–363.

NEW BOOKS RECEIVED

Arrighi, Giovanni, & Beverly J. Silver. *Chaos and Governance in the Modern World System*. Minneapolis, MN: University of Minnesota Press, 1999. Pp. x, 336. $22.95, paper.

Audi, Robert; General Editor. *The Cambridge Dictionary of Philosophy*. 2nd ed. New York: Cambridge University Press, 1999. Pp. xxxv, 1001. $74.95.

Babe, Robert E. *Canadian Communication Thought: Ten Foundational Writers*. Toronto: University of Toronto Press, 2000. Pp. x, 448.

Backhouse, Roger, & John Creedy, Eds. *From Classical Economics to the Theory of the Firm*. Northampton, MA: Edward Elgar, 1999. Pp. xiv, 306. $95.00.

Bazzoli, Laure. *L'economie politique de John R. Commons*. Paris: Editions L'Harmattan, 1999. Pp. 234.

Benjamin, Walter. *The Arcades Project*. Cambridge, MA: Harvard University Press, 1999. Pp. xiv, 1073. $39.95.

Besomi, Daniele. *The Making of Harrod's Dynamics*. New York: St. Martin's Press, 1999. Pp. xii, 289. $75.00.

Bethell, Tom. *The Noblest Triumph: Property and Prosperity Through the Ages*. New York: St. Martin's Griffin, 1999. Pp. vi, 378. $16.95, paper.

Blaug, Mark, Ed. *Who's Who in Economics*. 3rd ed. Northampton, MA: Edward Elgar, 1999. Pp. xx, 1237. $350.00.

Blaut, J. M. *Eight Eurocentric Historians*. New York: Guilford, 2000. Pp. xii, 227. $42, cloth; $22, paper.

Boettke, Peter J., Ed. *The Legacy of Friedrich von Hayek*. Vol. II. *Philosophy*. Northampton, MA: Edward Elgar. Pp. viii, 485.

Boggs, Carl. *The End of Politics: Corporate Power and the Decline of the Public Sphere*. New York: Guilford Press, 2000. Pp. x, 309. $23.95.

Bouckaert, Boudewijn, & Annette Godart-van deer Kroon, Eds. *Hayek Revisited*. Northampton, MA: Edward Elgar, 2000. Pp. xxi, 157. $80.00.

Buchanan, James M., & Richard A. Musgrave. *Public Finance and Public Choice: Two Contrasting Visions of the State*. Cambridge, MA: MIT Press, 1999. Pp. ix, 272. $27.50.

Busch, Lawrence. *The Eclipse of Morality: Science, State and Market.* New York: Aldine de Gruyter, 2000. Pp. ix, 219. $43.95, cloth; $21.95, paper.

Carmichael, Richard E. *Politics and Economics in America: The Way We Came to Be.* Pp. xiii, 369. Malabar, FL: Krieger Publishing Company, 1998. $29.50.

Cartwright, Nancy. *The Dappled World: A Study of the Boundaries of Science.* New York: Cambridge University Press, 1999. Pp. ix, 247. $54.95, cloth; $19.95, paper.

Cashmore, Ellis, & Chris Rojek, Eds. *Dictionary of Cultural Theorists.* London: Edward Arnold, 1999. Pp. x, 497. $19.95, paper.

Colander, David, Ed. *The Complexity Vision and the Teaching of Economics.* Northampton, MA: Edward Elgar, 2000. Pp. xv, 307. $95.00.

Cowen, Tyler. *In Praise of Commercial Culture.* Cambridge, MA: Harvard University Press, 1998. Pp. ix, 278. $14.95, paper.

Daniel, Sami; Philip Arestis, & John Grahl, Eds. *Regulation Strategies and Economic Policies.* Northampton, MA: Edward Elgar, 1999. Pp. xxxviii, 221. $95.00.

Diggins, John Patrick. *Thorstein Veblen: Theorist of the Leisure Class.* Princeton, NJ: Princeton University Press, 1999. Pp. xxxv, 261. $16.95, paper.

Emmett, Ross B., Ed. *"What Is Truth in Economics?"* Selected Essays by Frank H. Knight, Volume I. Chicago, IL: University of Chicago Press, 1999. Pp. xxiv, 406. $58.00.

Emmett, Ross B., Ed. *Laissez-Faire: Pro and Con.* Selected Essays by Frank H. Knight, Volume II. Chicago, IL: University of Chicago Press, 1999. Pp. vi, 459. $58.00.

Fanno, Marco. *A Contribution to the Theory of Supply at Joint Cost.* New York: St. Martin's Press, 1999. Pp. xlv, 131. $75.00.

Favretti, Rema Rossini; Giorgio Sandri, & Roberto Scazzieri, Eds. *Incommensurability and Translation: Kuhnian Perspectives on Scientific Communication and Theory Change.* Northampton, MA: Edward Elgar, 1999. Pp. xiii, 507. $100.00.

Feldman, Stephen M. *American Legal Thought from Premodernism to Postmodernism: An Intellectual Voyage.* New York: Oxford University Press, 2000. Pp. 272. $19.95, paper.

Finkelstein, Andrea. *Harmony and the Balance: An Intellectual History of Seventeenth-Century English Economic Thought.* Ann Arbor, MI: University of Michigan Press, 2000. Pp. ix, 380. $49.50.

Fleischacker, Samuel. *A Third Concept of Liberty: Judgment and Freedom in Kant and Adam Smith*. Princeton, NJ: Princeton University Press, 1999. Pp. iv, 446. $19.95, paper.

Fogel, Robert William. *The Fourth Great Awakening and the Future of Egalitarianism*. Chicago, IL: University of Chicago Press, 2000. Pp. 383. $25.00.

Frank, Dana. *Buy American: The Untold Story of Economic Nationalism*. Boston, MA: Beacon Press, 1999. Pp. xii, 316

Garnett, Robert F., Jr. *What Do Economists Know?: New Economics of Knowledge*. New York: Routledge, 1999. Pp. xii, 259. $19.99, paper.

Grassby, Richard. *The Idea of Capitalism Before the Industrial Revolution*. Lanham, MD: Rowman & Littlefield, 1999. Pp. ix, 85. Paper.

Greene, William H. *Econometric Analysis*. 4th ed. Upper Saddle River, NJ: Prentice-Hall, 2000. Pp. xxv, 1004.

Groenewegen, John, & Jack Vromen, Eds. *Institutions and the Evolution of Capitalism: Implications of Evolutionary Economics*. Northampton, MA: Edward Elgar, 1999. Pp. xi, 205. $80.99.

Heertje, Arnold, Ed. *The Makers of Modern Economics*. Vol. 4. Northampton, MA: Edward Elgar, 1999. Pp. vii, 174. $80.00.

Hutchison, Terence. *On the Methodology of Economics and the Formalist Revolution*. Northampton, MA: Edward Elgar 2000. Pp. viii, 383. $100.00.

Kabeer, Naila. *The Power to Choose: Bangladeshi Women and Labour Market Decisions in London and Dhaka*. New York: Verso, 2000. Pp. xvi, 464. $29.00.

Kenwood, A. G., & A. L. Loughheed. *The Growth of the International Economy, 1820–2000: An Introductory Text*. 4th ed. New York: Routledge, 1999. PP. xvi, 349. $90.00, cloth; $29.99, paper.

Kurz, Heinz D., Ed. *Critical Essays on Piero Sraffa's Legacy in Economics*. New York: Cambridge University Press, 2000. Pp. xiv, 458.

Kwass, Michael. *Privilege and the Politics of Taxation in Eighteenth-Century France*. New York: Cambridge University Press, 2000. Pp. xvii, 353. $69.95.

Layard, Richard. *Tackling Unemployment*. New York: St. Martin's Press, 1999. Pp. xii, 543. $79.95.

Lonitz, Henri, Ed. *Theodore W. Adorno and Walter Benjamin, The Complete Correspondence, 1928–1940*. Caambridge, MA: Harvard University Press, 1999. Pp. viii, 383. $39.95.

Louca, Francisco, & Mark Perlman, Eds. *Is Economics an Evolutionary Science? The Legacy of Thorstein Veblen.* Northampton, MA: Edward Elgar, 2000. Pp. ix, 234. $55.00.

Lutz, Mark A. *Economics for the Common Good: Two Centuries of Social Economic Thought in the Humanist Tradition.* New York: Routledge, 1999. Pp. xiv, 302. $18.99, paper.

Lyas, Colin. *Peter Winch.* Ann Arbor, MI: University of Michigan Press, 1999. Pp. viii, 216. $19.95, paper.

Lynch, Joseph M. *Negotiating the Constitution: The Earliest Debates Over Original Intent.* Ithaca, NY: Cornell University Press, 1999. Pp. x, 315. $42.50.

Martin, Michael. *Verstehen: The Uses of Understanding in Social Science.* New Brunswick, NJ: Transaction Publishers, 2000. Pp. x, 264.

Mohr, Ernst, Ed. *The Transfer of Economic Knowledge.* Northampton, MA: Edward Elgar, 1999. Pp. xxi, 214. $90.00.

Morgan, Mary S., & Margaret Morrison, Eds. *Models as Mediators: Perspectives on Natural and Social Science.* New York: Cambridge University Press, 1999. Pp. xi, 401. $64.95.

Morton, Rebecca B. *Methods and Models: A Guide to the Empirical Analysis of Formal Models in Political Science.* New York: Cambridge University Press, 1999. Pp. x, 326. $59.95, cloth; $22.95, paper.

Nimitz, Jr., August H. *Marx and Engels: Their Contribution to the Democratic Breakthrough.* Albany, NY: SUNY Press, 2000. Pp. xiii, 377. $23.95, paper.

Okumura, Hiroshi. *Corporate Capitalism in Japan.* New York: St. Martin's Press, 2000. Pp. lxi, 165. $65.00.

Parenti, Michael. *History as Theory.* San Francisco, CA: City Lights Publishers, 1999. Pp. xxi, 273. $26.95, cloth; $14.95, paper.

Peet, Richard; with Elaine Hartwick. *Theories of Development.* New York: Guilford, 1999. Pp. xii, 234. $22.95, Paper.

Pearl, Judea. Causality: *Models, Reasoning, & Inference.* New York: Cambridge University Press, 2000. Pp. vxi, 384. $39.95.

Perelman, Michael. *The Invention of Capitalism: Classical Political Economy and the Secret History of Primitive Accumulation.* Durham, NC: Duke University Press, 2000. Pp. 412. $64.95, cloth; $22.95, paper.

Persson, Karl Gunnar. *Grain Markets in Europe, 1500–1900: Integration and Deregulation.* New York: Cambridge University Press, 1999. Pp. xx, 173.

Pressman, Steven. *Fifty Major Economists.* New York: Routledge, 1999. Pp. xi, 207. Paper.

Pinkard, Terry. *Hegel: A Biography.* New York: Cambridge University Press, 2000. Pp. xx, 780. $39.95.

Prechel, Harland. *Big Business and the State: Historical Transitions and Corporate Transformation, 1880s–1990s.* Albany, NY: SUNY Press, 2000. Pp. xiv, 317. $25.95, paper.

Pyle, Andrew, Ed. Population: *Contemporary Responses to Adam Smith.* Bristol, U.K.: Theommes Press, 1994. Pp. xxv, 319. $24.00.

Reeder, John, Ed. *On Moral Sentiments: Contemporary Responses to Adam Smith.* Bristol, U.K.: Theommes Press, 1997. Pp. xxiv, 239. $22.00.

Rizzello, Salvatore. *The Economics of the Mind.* Northampton, MA: Edward Elgar, 1999. Pp. xxi, 198. $80.00.

Ross, Ian S., Ed. *On the Wealth of Nations: Contemporary Responses to Adam Smith.* Bristol, U.K.: Theommes Press, 1998. Pp. xxxviii, 248. $23.00.

Scheiber, Harry N., Ed. *The State and Freedom of Contract.* Stanford, CA: Stanford University Press, 1999. Pp. xi, 378. $55.00.

Sclar, Elliott D. *You Don't Always Get What You Pay For: The Economics of Privatization.* Ithaca, NY: Cornell University Press, 2000. Pp. xiii, 184. $25.00.

Seyoum, Belay; and Rebecca Abraham, Eds. *Sources: Notable Selections in Economics.* Guilford, CT: Dushkin/McGraw Hill, 1999. Pp. x, 367. Paper.

Shenhav, Yehouda. *Manufacturing Rationality: The Engineering Foundations of the Managerial Revolution.* New York: Oxford University Press, 1999. Pp. viii, 247. $49.95.

Silver, Morris, Ed. *Ancient Economy in Mythology East and West.* Savage, MD: Rowman & Littlefield, 1991. Pp. viii, 288.

Snowdon, Brian, & Howard R. Vane. *Conversations with Leading Economists: Interpreting Modern Macroeconomics.* Northamption, MA: Edward Elgar, 1999. Pp. xi, 370. $100.00.

Szenberg, Michael, Ed. *Passion and Craft: Economists at Work.* Ann Arbor, MI: University of Michigan Press, 1998. Pp. xvii, 314. $65.00, cloth; $23.95, paper.

Taylor, Mark C. *About Religion: Economics of Faith in Virtual Culture.* Chicago, IL: University of Chicago Press, 1999. Pp. 292. $19.00, paper.

Tindall, Gillian. *The Journey of Martin Nadaud: A Life and Turbulent Times.* London: Chatto & Windus, 1999. Pp. vii, 310.

Turner, Frederick. *Shakespeare's Twenty-First Century Economics: The Morality of Love and Money.* New York: Oxford University Press, 1999. Pp. viii, 223. $35.00.

Tuttle, Carolyn. *Hard at Work in Factories nd Mines: The Economics of Child Labor During the British Industrial Revolution*. Boulder, CO: Westview, 1999. Pp. ix, 308. $63.00.

van Creveld, Martin. *The Rise and Decline of the State*. New York: Cambridge University Press, 1999. Pp. viii, 439. $54.95.

Walker, Donald A., Ed. *Equilibrium*. 3 vols. Northampton, MA: Edward Elgar, 2000. Pp. xviii, 583; x, 598; x, 621. $640.00.

Wallerstein, Immanuel. *The End of the World As We Know It: Social Science for the Twenty-First Century*. Minneapolis, MN: University of Minnesota Press, 1999. Pp. ix, 277. $29.95.

Williams, David, Ed. *The Enlightenment*. New York: Cambridge University Press, 1999. Pp. xii, 529. $54.95.

Winks, Robin W., Ed. *The Oxford History of the British Empire: Historiography*. New York: Oxford University Press, 1999. Pp. xxiv, 731. $49.95.